Black

for

Breakfast

One River, Three Kayaks, Six Months

Hollie and Jamie Manuel

Create Space
Independent Publishing Platform

ISBN: Paperback first edition 978 1-5468-6534-6
ISBN: Ebook: Not Purchased

Biography & Autobiography/Adventurers and Explorers

This book is facilitated by the Createspace Independent Publishing Platform and is printed on demand by Amazon

To contact the author please email
mgogdesigns@gmail.com

Jamie Manuel lives for the wilds of Africa, for Mt. Kenya and The Aberdare Mountains. Inspired by the tales of George Adamson, Samuel Baker and Livingstone himself, he is determined to see as many of the remaining places of really wild Africa that he can. Armed with fearlessness and what can only be described as *more* than nine lives, he continues to get himself into sticky adventures from which he returns home with more unbelievable tales, photos, video footage and wounds.

Hollie Manuel (M'gog): 'I've always wanted to write a book but it's not easy to choose the subject of that first one. When it initially occurred to me that perhaps charting the careless adventures of my wanderlust brother might be an easy first attempt, I was wrong! There were sibling arguments, complete disbelief at many of the tales, fact checking that proved 'Brother Jim' was right and finally a long, hard slog to get all 300+ pages done ... and published.
'

For Dave:

Who nurtured and understood my wild love of the bush – I may not have found the Nandi Bear yet ... but I will!

And for Mugie ... a lion cub I will never forget.

Table of Contents

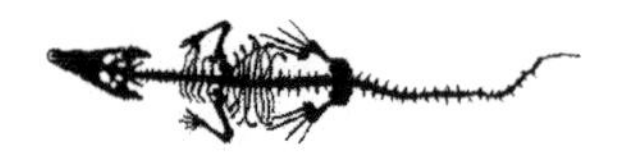

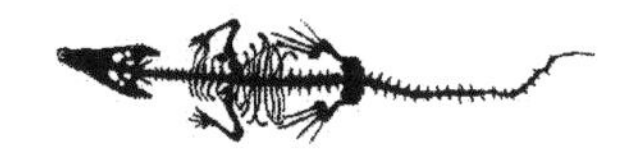

Maps & Sketch Inserts in order of view

Don't forget to look at our blog where you will find a link to the online version of the
COFFEE TABLE PHOTO BOOK

https://blackmambaforbreakfast.blogspot.com

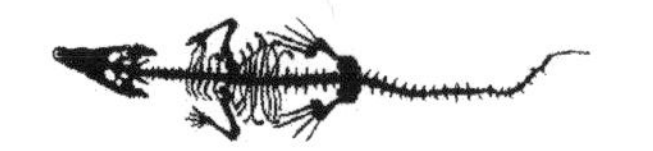

Acknowledging a Love Affair with a Lion

Every lion has a story, but my story and Mugie's are inextricably linked. I do not believe that destiny is planned, but I do believe that some souls are born to be wild and hardened by this world. I want to share with you the tale of my love affair with a lion called Mugie and how it was that he led me on to tell the adventures of my life thus far.

It is a short account, at the same time uplifting and heart-breaking because Mugie did not live long. He never became the grand maned king of the savanna; he did not even live long enough to lose the spotty undercarriage that he proudly groomed. Mugie lived only long enough to capture my heart and to highlight to me the importance of each of our lives – we *all* can make a difference.

Mugie came to me in 2011, a spotty, crying cub of three months, abandoned by his mother and siblings. The rains had come early and were heavy. The land had flooded and Mugie was trapped on a small hillock apart from his family, who had managed to reach safety on higher ground near-by. His mother, frightened by the still rising water and perhaps a first time mother, unsure of how to rescue her little one, mewed at him ceaselessly for days. But Mugie was too little to swim and eventually his mother knew that survival was cruel, and turned her

back on her smallest so as to save the rest of her litter. Luckily a friendly warden rescued the sodden lion-kitten and Mugie was flown up to me in Kora.

Kora had been the home of the famous lion Christian, watched by millions on You-tube when he greeted his former London owners after a year apart. We hoped Mugie would become a "Christian-Lion' in his own right: Would he have documentaries and films made about him? Would he walk the savannah as a proud king?

In the early days, when Mugie and I were getting to know one another, I fed him and he cried to me for food. We absorbed each other's smells and learned each other's boundaries – an important aspect of hand raising a lion. Mugie's paws grew faster than he did and often he would trip and face-plant into the dust, at other times he swatted at my legs and I had to reprimand him each time he unsheathed his claws. The hot, cloudless days flew by in colours of blue as both Mugie and I grew and matured into each other. Soon it was time to begin familiarising him with the arid thorn-scrub desert that lay beyond the *Kampi ya Simba* boundaries. I began to take him on twice daily walks. At first we stayed close by – we explored sandy *dongas* (dry river beds), we climbed the gently slanting rocky outcrops that are common in the area, and we chased sand-grouse and francolin. Mugie's soft paws began to harden up and as his strength grew he demanded longer and longer walks.

Then came Mama Ngina!

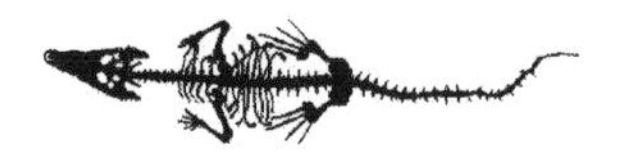

Mama Ngina was an ostrich chick that we had rescued and raised into a proud feathered bird. Generally Mama Ngina lived a few kilometres away at the airstrip, but as soon as she got to know there was a lion next door, she was hard to turn away! And so the early morning and eventide walks became a family affair with ostrich and lion chasing each other, cavorting between the thorny *Commiphoras*, playing hide-and-seek about me. I was careful to be as hands off with Mugie as I could, after all his destiny was reintroduction, and without a fear of humans he could become a problem animal. No camp visitors could touch or pet Mugie, and only I could feed or walk him. Generally, he ate donkey meat that was skinned and thawed for him each day, but soon he became adept at catching wild birds and an occasional baboon. Baboons were always a constant threat to a young lion and we suffered large mob attacks that always ended with gunfire as I was forced to scare away the large fierce males who came too close for comfort with teeth-baring, loud barking attacks that terrified both Mugie and me.

I began the basics of teaching Mugie to hunt through dragging lures across the ground, up trees and about boulders. One dusky evening, as the francolin called from the golden sands and the owls ruffled feathers ready for their night forays, Mugie tried his first roar ... a guttural half squeak after which he puffed out his chest and tore up my flip-flop.

Arid days followed arid days and the sun relentlessly dried the ground as we waited for

our first rains. Nimbus's of cloud would pile up and then melt away on an ever tantalising horizon. Even Mugie would raise his whiskered muzzle to the sky and search for rain.

Rain had now not fallen for two and a half years but one grey, windy beautiful day it broke and Mugie stood nose sky wards, whiskers dripping. I stood in my own delight, arms wide open, the rain washing off me and into the thirsty sand below my feet. Mama Ngina twirled with me as the smell of fresh rain scented the air and freshened the year. Clouds were torn open by lightning that electrified the rock kopjes and thunder bellowed warnings at all to take cover. Mugie seemed delighted, we were now many miles from camp, and he and I headed for a cave up in the hills and like two old friends, we sat in the cave looking out onto the African wild-lands as they received the gift of rain.

The very next morning new green shoots were pushing their way up from the damp earth, the *Commiphora* was haloed in green and the birds sang as if it were the Garden of Eden. Elephants began to come back into the Kora National Park and with them poachers. The late George Adamson of *Born Free* fame had been murdered between here and the airstrip (Mama Ngina's home route) in 1985 by poachers or *shifta,* as the Kenyan-Somali bandits are called up here in the lawless wild west of the Northern Frontier District (although technically Kora lies at the heart of Kenya it is often classed as part of the NFD). These bandits make a living from elephant ivory and would stop at nothing to get

their prizes. The last rhino had been slaughtered in the late 70s.

Part of my job involved tracking the poachers and counting kills from the air in my tiny, two seater plane. 5Y KSK, the oldest flying plane in East Africa, made in 1950 and crashed many times by each of its many owners. Each day my heart sank lower as I counted ten – eleven – twelve fresh carcases. Wise old elephants with their ivory hacked off with machetes and their meat left to rot. Unnatural numbers of hyenas collected around the carcases and formed blood-hungry hunting clans that were scared of nothing.

We had to watch Mugie carefully as a young lion at a year and a half made a fine target for the marauding clans of loping hyenas. On walks now I was armed and often had to fire shots into the air to keep away the hyenas. At night we would lock Mugie in his large enclosure and the hyenas would come to laugh at the wire and slink along the outskirts, while Mugie, still not confident in himself, roared haltingly back at them.

At the changeover of the seasons Mugie had taken to spending nights away from camp. He would arrive back in the morning tired and panting and would collapse into a deep sleep of rich dreams, his paws twitching as the chases played out across his eyelids. It worried me that he might stay out before he was equipped to deal with the numbers of hyenas that a wild lion would not normally find in nature. This was a man-disrupted ecosystem that was off-kilter and dangerous. Even for me. Camp had to be

patrolled at night and twice we had to 'evacuate' and spend the darkness hours in the bush, hiding from the poachers whose shots rang out, too close for comfort.

One night I heard sounds that strangled my breath. Mugie had not come home that evening and I had stayed up awaiting his late return. Eventually, beneath the midnight moon, I had fallen into a troubled sleep. And now I was awakened by screams that lurched across the night. Hyenas. Mingled amongst the lilting half giggles and rasping coughs was the sound of Mugie!

'Mugie!' I yelled, already dressed and lurching unsteadily from sleep towards the car.

I fired off two shots and momentarily the fight stopped, only to renew more vigorously, more ferociously. It seemed to be some distance away, perhaps down at 'The Graves' (Terence, George Adamson's and Boy, George's lion). I climbed into my little yellow Suzuki and turned the key. Nothing. I could hear Mugie moaning and my palms were sweating as I tried again. Nothing.

'Damn it!' the car was so unreliable and temperamental. It took five or six minutes to get the car started and another three to reach Mugie.

The scene was horrific. Tufts of bloody hair lay clotting on the road. Five hyenas had set upon Mugie and he had valiantly fought them. When I reached the scene Mugie was frantically clawing the trunk of a small *Combretum* to pull himself up and out of harm's way, but his strength had ebbed and the hyenas

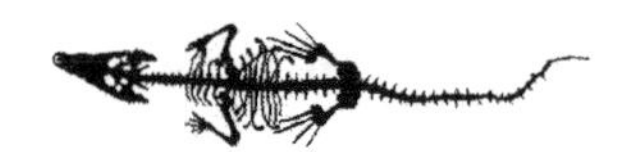

were stronger than him. I fired shots and chased the hyenas away, but they circled, darting in again and again to try to finish the kill. My Mugie. To kill my Mugie. Eventually they relented and I was able to get out of the car and to Mugie's side. The pads on his paws had been ripped off and bloody claws protruded, blood and sand bathed the wounds all down his side and his testicles were ripped and hanging. Mugie knew who I was and attempted to get to me, panting and groaning in agony. He used his last ounces of strength to pull himself to his feet and to haltingly walk some twenty yards before he collapsed. The guys from camp appeared and we built a circle of fires around Mugie, I drove the couple of kilometres back to camp to send out an alert on the radio. I knew that a wildlife vet from the David Sheldrick Wildlife Trust was in the next door National Park, Meru, to pick up an elephant orphan.

'Meru, Meru this is *Kampi Simba*, Come In.'

'Reading *Kampi Simba*, how can we help?'

'Mugie's been attached by hyenas, we need the vet ASAP.'

'Noted. Vet is available only in the morning, can you fly in and pick him up from here?'

'Affirmative.'

'Will Mugie hold on?'

'I hope so. Please ... as early as possible. Over and out.'

I rushed back to Mugie's side and we spent the next four hours monitoring his heart

rate and muttering confused words. It was all we could do. Come dawn I drove to the airstrip flew to Meru and picked up the vet. There was no need for words, my face held the story of the night, my clothes reeked of blood and fear, and my nails were bleeding and black.

We sedated Mugie and treated his wounds as best we could. He was castrated and the wound stitched up with tiny, offensive lines of thread; a king no longer. His paws were rinsed in antiseptic and bandaged and two long-lasting antibiotic shots given before we carried him gently back to camp, and to his little lean-to den in his enclosure. I was left with another shot of antibiotic as the vet took to the skies in the light aircraft that had come to collect him.

For the next thirty-six hours my heart bled and my mind dulled as I watched Mugie for any sign of recovery. I remembered those evenings where we had wrestled beneath the setting sun on the warm rocks on Kora. I recalled those moments when I was teaching him to climb trees where he would look comically up at me wondering if this really was lion behaviour.

We made sure that Mugie would not become dehydrated by attaching a liquids drip that I refilled several times. Finally, on the second morning as I sat watching the sun rise, Mugie poked his head out from his den and flicked his ears. He gave a small, soft mew and lay down again head on his paws. Tears welled up and ran like the recent rains down my cheeks. I moved to be with him and lay down beside him. Mugie closed his eyes and lapsed

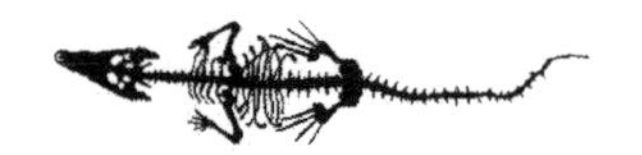

into death. That was it. Like a soldier in battle holding his dead brother, I kissed him for the last time, my heart broken by the love of two souls who would never again be able to walk in the wild together. I was torn; my life was broken into pieces, my finest friend was dead.

That was the end of Mugie's story. Of Mugie's lion life.

Now, six months later I have at last found it within myself to re-visit his grave, which lies next to those of George and Terence Adamson. I have re-walked the scene of the attack and have stared deep into the claw marks that scar the *Combretum* tree into which Mugie tried to climb to safety, and I have shed more tears over the ashes of the fires that we lit around Mugie to keep the hyenas away. It was here, as I remembered Mugie and his warmth that my mind flicked back to the book that my sister had suggested we write. *Maybe*, I thought, *maybe writing would capture all of the things that I love and remember, all of those memories that are so easily lost. Mugie was a story, but his life was too short to make a story of any length, instead he could be my inspiration to write.*

Mugie, I promised, *you will be the beginning of my memoir, a memoir that will flow like a river through all of my adventures. You shall be my source. My adventures shall intertwine with our tale of the Zambezi.*

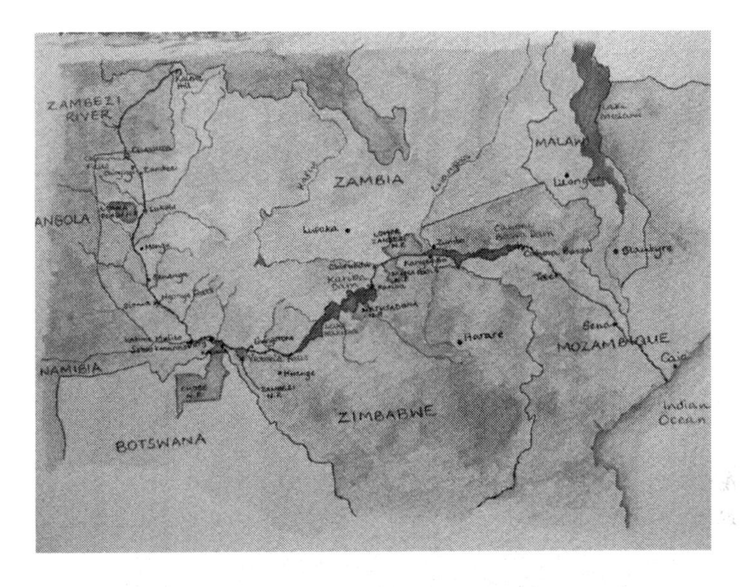

The Zambezi River flows for over 2,500 kilometres from a spring in the Zambian wetlands (the sloppy country) to the Indian Ocean. The Zambezi is Africa's fourth longest river after the Nile, Congo and Niger. Dr. Livingstone believed it could be a highway into central Africa from the Indian Ocean but the Cahora Bassa rapids in Mozambique dashed that dream.

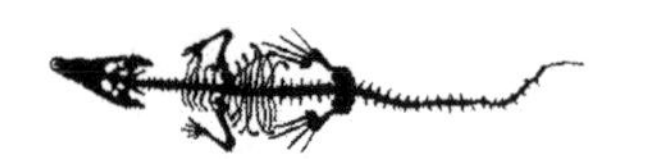

Part One

In the Beginning there was Imire

To me, Imire was the word,
And the word was Imire, for
'What is a man without the beasts?
If all the beasts were gone, man would
die from great loneliness of spirit. For
what happens to the beasts, soon happens
to man. All things are connected.'
Chief Seattle

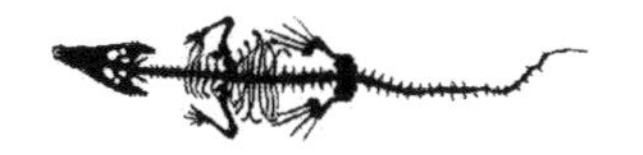

Chapter One

Broken In By Africa

Honour Africa as though she is
your father and your mother

Africa broke me. She thrust aside my ideas and values and made a mockery of them. She unwrapped my cocoon of safety and pitched me into the unknown. But the unknown is what I thrive on. I love the danger, the disease, and the desperation. Africa broke me, but she took my raw elements and built me up again - alive and inspired. I am who I am because of Africa and because of Imire, Reilly's home in Zimbabwe. I still have so much more to learn. And Reilly is no different. Allow me an introduction ...

Like a child caught thieving, Reilly caught my eye.

'Hey guys, check out my crocodile roll,' Reilly shouted, then flipped himself upside down.

Ace and I looked. Nothing happened. Ace and I stared. Still nothing happened. Finally bubbles surrounded a frantically rocking boat and a coughing and choking Reilly surfaced, red-faced and furious.

'What happened boys? Why didn't you rescue me? You plonkers!'

'You told us you were doing a crocodile roll. You didn't tell us to rescue you.'

'I know but if someone is drowning you rescue them.'

'But you weren't drowning. In fact you weren't even moving.'

'Cause I was dead.' Reilly burped a lungful of

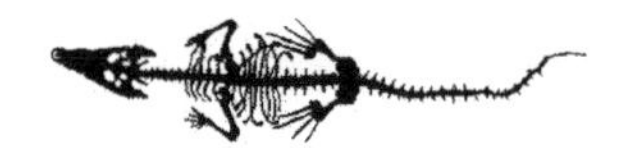

water out onto the pool's edge. 'Idiots!' he fumed.

This was our third day of pool session practice and we were not getting anywhere. Not one of us could roll.

'Listen you *oakes*, let's go down to Vic Falls and practice there. There's bound to be some dude who can teach us how to roll' Ace suggested. 'And we can have a party.'

Reilly snapped out of his mood and we notched in a weekend.

Reilly was my closest friend. My partner in crime. He came from Imire Safari Ranch. A small wildlife and cattle ranch in Hwedza, Zimbabwe. It was his tales of the ranch that he told that brought me to Zimbabwe. I met Reilly at Bush Academy in South Africa and followed him home when our year at the academy was done. It was there that I discovered the off-kilter magic of Imire. Imire became home for four years of my life, with John and Judy my parents away from home and Reilly my brother in crime. Imire was the sound that my heart made as it beat loud and clear within me. Nothing on Imire surprised me because I had heard all the stories from Reilly at Bush Academy.

Pick-up loaded, Reilly sprawled on top of the kayaks, in a state of intoxication from the night before, and Ace and I in the front, we set off in high spirits. But the trip was a comic fiasco from start to finish. Reilly managed to doze off and the delinquent wind cooled his skin as the furious sun partook in the game to scorch him carmine!

Ace and I practised hard with our boats fully loaded as they would be on the expedition.

'You *oakes* are hopeless,' old mate Victor, our ever patient kayak instructor, said. 'First off, it's not called a croc roll, it's an Eskimo roll. And second off,

with those kayaks fully loaded you're not going to be able to roll back up however good you are. Pull out now. Those river crocs, they gonna eat you for breakfast.'

As if to highlight the sheer hopelessness of our efforts, the baboons screamed their mirth from the trees and then ran off with some of our expedition supplies and two pairs of shoes. Reilly managed to ricochet a golf ball off a cliff that came right back at him and smacked him in the nuts and Ace managed to eat a load of dog food in a drunken state, and made himself very sick. Eskimo rolling practice was clearly out. The tourist activities of the falls were in, and clearly more appealing.

'And you guys are from where?' the white water raft guide asked, his *Zimbo* twang obvious.

'From Zim. We have come down to look at the rapids. We want to run them later this year in kayaks.' The raft guide raised first one eye-brow, and then the second.

'Then I take it you are experienced kayakers?'

'Uh no,' Reilly replied, 'do we need to be?'

'Well, you ain't gonna get down those things if you're not,' he said with finality.

With helmets firmly fastened, paddles resolutely gripped and our wits gathered we plunged into the first rapid.

'Left paddle,' shouted the raft guide as the water roared around us 'right paddle ...and...Get down!'

We all crouched inside the raft, knees at our chins, eyes closed. Again and again one or another of us was flung from the raft into the tumultuous water. Water that sucked us into its fizzing depths and spat us out downstream where a safety kayaker paddled furiously, with us clinging tightly to his dinky craft, to take us out of the current and into the eddy

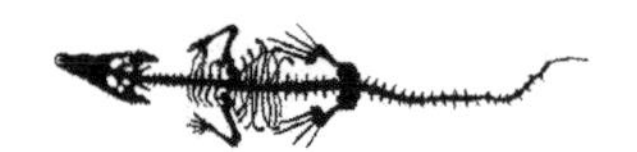

so that we could re-board the raft again.

'Well boys, what do you think?' the raft guide asked as we reached the top of the gorge, blood pounding in our ears 'you gonna do that in kayaks?'

'Uuuh' was all Ace and I could say. Reilly said nothing. His earlier bravado had floated down the river with the white-capped rapids.

Next on the adrenaline to-do list was the bungee jump.

'If you can do the spit test you're fine,' the man in charge told me.

I could. My saliva had not quite dried up yet. The stone platform with its wooden ramp stretched out into yawning space. My harness felt tight around my hips.

'3 ... 2 ... 1 ... Jump!'

I jumped. The sheer cliffs on the other side of the Batoka Gorge seemed to rush towards me. I distinguished a thousand hues of brown before I managed a breath. Below me the turbulent white water seemed harmless. The rich green figs clung to the vertical stratum, and far beyond me the marbled water rushed out of sight on its journey to the ocean. A journey we too would make.

Those honeyed days that we spent at Vic Falls gave us none of the preparation that we had hoped for but, despite that, we concluded on the long journey home - Reilly now in the front with us to save his ass from any further exposure to the laughing sun - the expedition was still possible.

'Of course,' Reilly grinned, 'we just have to take our time.'

In its essence this is not a tale of a well-planned, well-executed expedition. Nor is it a story about three especially good kayakers. In fact we had never been in a kayak before training for this expedition. This is an intricately woven story with

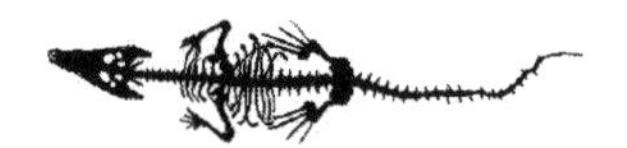

threads of a river tale, linked with threads that come from past memories and adventures that I have been lucky enough to fall into. This is a story of three people with a deep love for Africa and her wilderness, who concocted a grand idea, acted on it and succeeded, (against many people's better judgement).

Reilly and Ace are from Zimbabwe and I am from Kenya. All of us are second or third generation Africans originally from farming families. Reilly and I were Bush Academy graduates; we lived, studied and worked together at the FGASA (Field Guides of Southern Africa) guiding school in Leadwood National Park in South Africa. Later I moved to Zimbabwe and worked with Reilly on the game farm that his grand-parents and parents before him had built up. Imire Safari Ranch is home to one of the only black rhino breeding centres in Africa. The ranch is also home to five tame elephants that came from troubled backgrounds, buffalo, giraffe, eland, sable and many more. It was here, in 2007, where I met Ace. He lived in Harare and was a friend of Reilly's.

Our Zambezi expedition broke no records. It was not even an unusual idea to begin with. Many a young adventurer has dreamed of completing a river, any river, in a source-to-sea expedition. If pushed, any other person could have completed it in a far shorter time. And they have. Our only audacity was to go against the advice of all who knew the river and its dangers: the crocodiles and hippos, the rapids and whirlpools, malaria and snake bites, sunstroke and dysentery. We paddled with no time lines. Distance bore no relation to the times we took. We were three young adventurers with Africa in our hearts.

Although we completed as much of the source-to-sea of the Zambezi as we could, we were

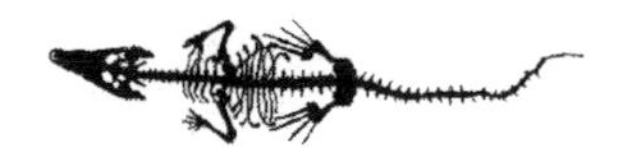

not able to paddle the Angola section (two hundred and forty kilometres) as there were security issues in the area at the time. We had hoped to return and complete the section later, but time has moved on. Ace and Rei are now both married, Reilly is back at Imire Safari Ranch expecting a child, and I have moved on after having worked for the George Adamson Wildlife and Preservation Trust up at Kora National Park in Kenya, starting back where the late George Adamson left off. Twenty three years after George's murder I was the keeper of these crazy-beautiful wild lands.

Preservation of Africa's wildlife will always be at the forefront of my consciousness, and I will do all I can towards its protection. Up in Kora my time was spent doing just this, as the new team tried to re-open the park to fulfil a 2030 vision in which tourism will be the park's life-blood. But, for now at least, we have had to retreat as presently this wild and harsh park is dangerous and lawless. Living there is precarious because the fight these days is with elephant poachers, aid money, hungry pastoralists and above all – population growth.

From there I moved to work with the Northern Rangelands Trust to run their anti-poaching operation and to train rangers as well as oversee their new rhino sanctuary up in Northern Kenya.

Conservation is Africa's newest battle and until Africa's population is brought under control I believe it is a battle we are sure to lose, a battle though that I will continue to play my part in..

But why risk our lives to run the Zambezi all in the name of conservation you may ask? Why chase an adventure that may kill us?

The answer for me was Imire.

Chapter Two

Learning to Live Wild

Thou shalt make no covenant with poachers, but work to bring them from the dark side

We accepted an invite from John and Judy, Rei's parents, for dinner.

'Just beware Tsotsi,' Judy warned us, as she did each time we came for dinner. Tsotsi was the pet hyena. Zazu was the family dachshund (dacksi) who thought she was a hyena.

Imire had three sections to it: The tobacco faming side and John and Judy's house, the lodge with Nzou the elephant and buffalo and Numwa, our section, which housed the wildlife ranch and the volunteer project set up by Reilly and I.

Arriving for dinner in the Land-Rover posed no problems as we could drive right to the door and discharge guests relatively easily, but on the motorbike *Tsotsi* posed more of a problem.

'Rei, here he comes!'

'Hold on, I'm gunning the engine!' Rei would grit his teeth and loop the house at full speed, *Tsotsi* snapping at our rear wheel.

'Check the window, Mum!'

Judy's face would appear momentarily at the window we needed. 'All clear Rei, just don't smash the glass.'

We had done this so many times now that we had the sequence down pat. As Rei gunned the engine, keeping just ahead of *Tsotsi*, I was already sitting on my haunches on the seat behind him and as he rode past the conveniently low and open window I would leap from the bike and through the window to safety while Rei did another circuit. As Rei came round again he too had his feet on the saddle and was using only the throttle to keep his speed. I quickly moved as Rei came flying through the window and the motorbike roared off on its own into a soft buffer that we had made for exactly this reason ... the engine spluttered and died and Rei and I hung out the window to shout some abuse at *Tsotsi*.

'Beerboys?' Judy ran her words together as she called from the kitchen. She was entirely used to her children appearing home in wild ways and nothing fazed her anymore.

'Yes please.'

Sam, Rei's youngest sister, was home and she too was not too friendly with *Tsotsi* – in fact she and *Tsotsi* were sworn enemies and Sam had to carry a great big stick with her when she was home. Although *Tsotsi* was locked out now while we had dinner, generally he had run of the house.

'Bloody hyena,' Sam fumed, her freckles dancing with rage. 'I think Mum and Dad love *Tsotsi* more than me!'

'Not really,' Judy shouted from the kitchen, 'but, Sam, you are always away at boarding school so you can't complain about *Tsotsi* in the house just 'cause you are home.'

'Can too!' Sam shouted back with a smile.

Judy appeared from round the white wall that sported muddy paw prints of all sorts of animals. 'Mum you even let him live under my bed!'

Judy had found *Tsotsi* abandoned on the farm and had not been able to resist taking him in. As Sam was away at school *Tsotsi* made his den under her bed and when she came back for holidays and forcibly removed him, a life-long enemy was made. *Tsotsi* had now reached a size and jaw strength that was alarming but John and Judy had lost their hearts to him. Now he ranged freely around the house and garden and frightened everyone he came across. Only Judy's little dacksi, Zazu could control *Tsotsi*.

Woof woof woof she would bark and *Tsotsi* would rush out into the garden whilst Zazu quickly gobbled down the meat from Tsotsi's plate. *Grrrr* she would growl if *Tsotsi* tried to usurp or even share her warm space in John and Judy's big double bed in the morning.

But hyenas and dacksis, along with an elephant who thought she was a buffalo were the least of the worries. Tatenda also shared this great big, thatched farmhouse with its colourful flowerbeds and deep blue swimming pool ringed with lush green lawn. Tatenda felt that he fitted in fine amongst the chickens, the cats and the dogs, but really he had to squeeze between the living room furniture, and when he laid his great big head upon the bed each morning his happy squeaks made John and Judy query the future. Tatenda was a large six month old rhino orphaned when his mother was shot dead by poachers on the game park side of the farm.

'Tendy,' Judy crooned, 'oh Tendy, what are

we going to do?' and furrows appeared on John's brow and Reilly bowed his head as we all remembered the awful night that left Tatenda orphaned.

It had happened just after dark one evening, after we had carefully locked and penned the herd of black rhino, and had settled in for the night, Lizzie our house-help breathlessly appeared at the window.

'*Baas* come quick! Poachers they have come, they are killing the rhino and they have tied us up, but me I have escaped!'

We heard shots but were already reaching for our weapons. Reilly took the wheel and we rattled the old Series Two Land-Rover, as fast as we could go, the two kilometres to the rhino *bomas*. As we drew close we fired our own shots into the blackness of the night. There was no come-back.

We released the staff that'd been tied up but sadly the poachers had already scarpered.

The scene was horrific. DJ's face had been hacked apart in the poachers' quest for a half inch stub of horn, all that remained on the already dehorned rhinos. Amber, who had been in calf and due to give birth in a month, had been gunned down and we were not able to save her calf. Noddy, a male, was also lost in the blood-bath that night. Only Tatenda, a four week old rhino, managed to survive. Tatenda was deeply in shock and it was vital we made him warm and gave him fluids now that his mother was no longer there to feed him. The scene on this day would leave an indelible print on my mind's eye. The devastation I felt to see the magnificent beasts that I had worked with, side-by-side, year after year, slaughtered and hacked apart on the red African soil, drove a spear deep into my heart.

Nine tonnes of magnificence ... gone ... in less than nine seconds.

These rhinos were our family and we knelt for long minutes cradling their huge heads, tears running down our faces as if the rain were trying to wash away earth in a storm.

Tatenda was a tiny legend who lost his entire family that night. He was taken from the carnage and was hand raised to make a fabulous recovery before being re-introduced back into the farm rhino herd where he has now sired a calf of his own. He even starred in his own film in 2009, *There's a rhino in my house,* which showed Imire and its constant struggle in a country that was falling to pieces.

I though back to the time before I had ever visited Imire and Rei's tales that so enraptured me.
'I love Imire,' Reilly would whisper and then; pause ... 'and you will too. We also have an elephant who thinks she is a buffalo and a warthog who thinks she is a dog. And that is an aside from Zazu who thinks she is a hyena!'

Reilly loved telling campfire tales at Bush Academy, and we all loved listening. His home seemed enchanted.

As he swilled a sneaky whiskey round in his glass, an amber fire would break out and engulf his face. His words crackled like faulty-circuit fireflies on the air.

'Nzou, she's a fine elephant,' he would say in a voice that belied his twenty years. 'Imagine, Jim, they were going to shoot her. Dead. Gone.' Rei looked up at the stars and sighed. 'Ahhh. You *know* I want to see Imire leading the world in conservation. I know we can do it, and we'll share. We'll show all those who want to see what it is like to live in a part of Africa where we care.'

'And Nzou?' I prompted.

'Nzou. Yeah, well. Hwange Elephant Park was

going to shoot her. Well, Mum decided that we would take her on. Crazy at the time, I remember. We darted her, loaded her onto the flatbed truck and drove her down here. Thing is,' Rei paused, 'we had no elephant on the farm so we didn't have a clue in hell where we were going to put her. I remember Gramps saying 'well, we got the buffalo Rei – maybe she'll take to them.'

I'd learnt all about the breeding of foot-and-mouth/theileriosis free game for sale to other ranches and at auctions. These buffalo fetched higher prices and ranches could earn a good name against their breeding herds. Would it be wise to introduce an elephant?

'I tell you it was mayhem,' Rei continued, 'when Nzou arrived we reversed the M99 sedative and just plain released her. Jimbo, its amazing man, you have to come visit one day.'

And less than six months later here we were – Nzou grazed in front of us amongst *her* herd of twenty seven buffalo. She had readily accepted her role as matriarch of the buffalo herd, much to the chagrin of the big male buffalo who were not used to being led by a lady. And a lady is what Nzou was and is. Instinct probably told them that things were askew but then, it's never wise to challenge a colossal beast. The buffalo had changed their habits accordingly. They followed Nzou wherever she led them, they answered her trumpets when she called and they permitted her fondling throughout the day.

One specific behaviour though, had changed hugely. Normally a female would calve with the herd around her, numbers mitigating the chance of predators pouncing. Now though, the female buffalo in this herd, Nzou's herd, disappeared into the bush when birthing. Nzou loved her charges but seemed not to be able to recognise the smell of a newborn

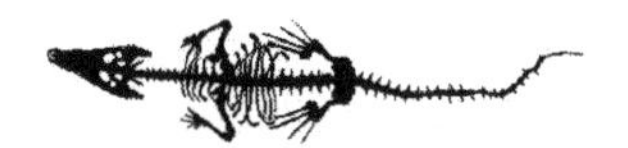

buffalo; as a result there had been several cases of trampling. Now the buffalo had learnt, and for a week or so they kept their newborns away. When they *did* return Nzou checked them over with her be-whiskered, wrinkled grey trunk and accepted them cordially into her herd.

'The trouble comes when the males reach maturity,' Rei was saying as he chewed on a grass stem, his tanned arm hanging lazily over the door of the old Land-Rover. 'Nzou knows that a big male buffalo may challenge her, and so she takes matters upon herself and kills them once they reach maturity. We have to manage the herd carefully and remove any potentially competitive males to our bachelor herd on the other side of the farm. System works well now we have it figured.'

Imire didn't work like other game farms I had been on. Here Nzou had a keeper, a wizened old man called Mutambwewa. In 1994 Mutambwewa had been gored by a male buffalo. Nzou had charged to the rescue and the commotion was heard from the lodge. Nzou cradled her keeper in her trunk as the buffalo tried to circle in for a second goring. Carrying him gently to the fence-line Nzou deposited him on the other side and then spun in rage back towards the offender. Within five minutes the death sentence had been dealt. Mutambwewa couldn't even take too many days holiday because Nzou wouldn't accept a new keeper, she would only tolerate him.

Mutambwewa gave us a toothless grin and reached out with his gnarled claw of a hand to give Nzou a sharp slap on the knee as she reached into the car. We sat and absorbed the moment in the shadow of the massive grey beast that was twice the size of our little red Land-Rover. Rei shook his sheepskin mane of hair and laughed heartily as he pulled himself up and dug deep into his pocket.

'She can tell I've got oranges hidden!' Nzou gobbled down the orange happily and as juice dribbled from her hairy lip she burped an orange-flavoured burp.

'Charming, Nzou old girl, charming.'

Nzou turned her great rump to us and wandered into her herd of buffalo.

At that moment Judy, Rei's mum, pulled up on her motorbike with her little black dacksi, Zazu riding front feet up on the handlebars. Judy flashed her enormously wide, tanned smile from beneath her enormous straw hat, her hair peering out in tufts from beneath.

'Morningboys!' Judy again ran her words together and relayed them with an enthusiasm that, in times to come, would catch us off balance if we were reeling from hangovers. 'Jimbo, you getting the farm tour are you?' Her accent was typically white Zimbabwean and the words swayed against one another.

'Indeed, Judy. Rei's showing me the ropes.'

Time had swallowed days and months whole, and now I had been on Imire for several years. I knew Nzou intimately and had explored every thorny corner of the farm and game park. Rei and I had worked hard to set up a volunteer organisation, primarily to afford us an employment of sorts, and a wage, but also to bring in foreign currency. Zimbabwe was scraping its bottom on the rocks. The infamous war vets were at Imire's boundary fences making demands and threatening staff. Food prices had more than trebled and continued to rise; inflation made Zimbabwean bank notes almost obsolete (although we did light a few back-burns in fire-fighting season with them). On our side of the farm Rei and I went back to bartering. Surprisingly, volunteers did come despite the travel warnings, and

to feed them we tested every route we could. Meals at Imire were a gamble.

Living at Imire was a gamble.

John lamented daily. 'We're almost ready to bring in the tobacco crop and that's what these bloody war vets are waiting for, so next month we need all security stations on alert.'

'War vets' was the name given to the thug like youths who led the Mugabe approved invasions of white owned commercial farms. These youths were too young to have participated in the independence war but they leached off a previous threat made by the ZNLWVA (Zimbabwe National Liberation War Veterans Association) to the British Government. The ZNLWVA vented the frustrations of landless veterans and blamed Zimbabwe's white minority of predominantly British descent, for refusing to participate in constructive land reform. ZNLWVA threatened a 'bloodbath' in 'future clashes against commercial farmers' unless land hunger was addressed to satisfaction. President Mugabe allowed these violent and lethal invasions to occur and it was these that Imire was up against.

Staff meetings were tense and all options were considered. Things had reached the stage that Rei and I could not take the same route to the head farm office from our volunteer station on Numwa now. Death threats were common place and the war vets were trying their damnedest to get what they could. But of course they didn't want to work to earn, they wanted only to take. After all, what was the point in taking over a farm when the tobacco crop was still in the field with work to be done? Better to take the farm when the tobacco was cured and ready to go to market.

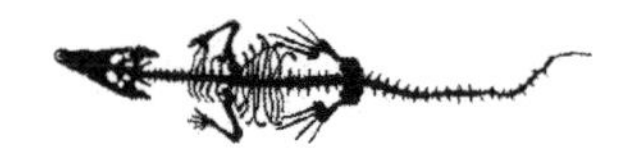

'Jamie and Rei, we need all eyes on the rhino. You'll have to take turns on night patrol, eleven p.m. and three a.m. please but not too scheduled, we don't want people guessing patrol times.'

Judy interrupted, 'and boys I hear there is very little food on the market at the moment so you are going to have to work hard at your veggie garden because it might need to support us for a while to come.'

Back at Numwa House, the volunteer base, we shovelled elephant manure by the truckload and made deep, rich soil mixes that we built up into seed beds over which we rigged shade netting to keep away the birds. Volunteers walked with the rhinos, recording their browse species and documenting their behaviour while Kutanga the baby elephant mock charged the volunteers as he wandered with his five strong herd beneath the *msasa* trees. Even the chickens were wary as Kutanga's favourite game was chasing them helter-skelter as he practised his trumpeting.

Enter Pog.

If a bite-happy hyena, a playful elephant calf and the often temperamental rhinos did not give the volunteers a run for their money then Pog did. Pog was a wilful female warthog who had been orphaned and then rescued and reared by Judy and handed over to us in the Numwa game park section. Pog's moods swung with very little grace first one way and then the other.

'It's not her fault Jimbo,' Judy said, taking the side of Pog as she did with every animal. 'I think she is pregnant.'

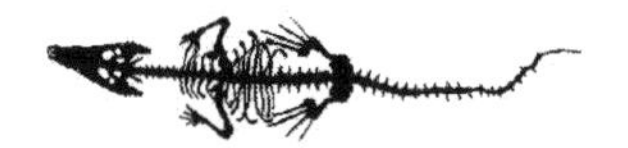

Sure enough Pog fattened up, and as her belly swelled so her moods grew worse.

'Rei we have to get her into a burrow somewhere, we can't have her babies running around anywhere and everywhere.'

We were not quite sure if Pog knew she was a warthog, after all she had been reared with dogs, a hyena and a rhino, and in fact her and Tatenda, the rhino, were 'bestest' of friends and even shared a great big double hay-bed. On a cold morning Tendy and Pog would be side by side, snuggled up in the hay and coaxed awake only by the promise of a big tub of milk that they snorted down together.

And so we began walking Pog to and from a burrow that we had found and bettered some eight hundred metres from the volunteer house. Often we threw feed-cubes down the tunnel to coax her down. Day by day she grew more and more used to the burrow and its confines and one day she inserted herself, rear end first, and refused to come out. We left her to it and for the next week we caught only glimpses of her as she dashed down the opening as we approached. She wanted no socialising. The volunteer house and Montague, the volunteer dog, a huge Rhodesian ridge-back could claim back their territory. A ceasefire had been called between the running battles that he and Pog fought daily.

As bronzed leaves fell from the *msasas* and the black coats of the sable began to lose their summer sheen, the air filled with the acrid smells of the Matabele ant and occasionally with the burnt, cheap cigar-like smell of bush fires.

Imire filled my mind and my body. At night we fell exhausted into our beds, still smelling of the

day's work if we hadn't made the cold waters of the dam for a wash before sleep.

Summers were golden and green, smelling of fresh water and elephant poo. As the summers faded into autumn the smells changed to that of dry earth and lightning littered the skies and often struck close. Winters were cold and misty, hot dry days that made my skin crack and my toes wrinkle, and cold nights that found us all snatching extra blankets from the empty volunteer beds. Spring brought promise but the promise was not silver lined. Still the war vets clamoured to be given land for free. One of John's new tractors was commandeered by an exceptionally brazen horde and taken to a farm some miles away, where it sits to this day, unused and rusting.

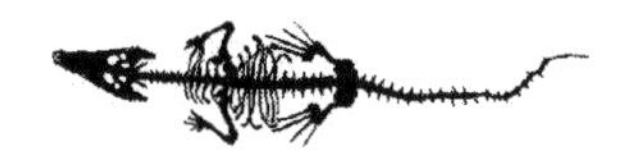

Part Two

Prepping Paddles

The Nile crocodile possesses sensory pits in the scales along the side of its jaw. These can detect movement and vibrations in the water at a distance of four hundred metres.

Fact.

It was fatalism with a loophole, and all you had to do to make it work was never miss a sign. Survival by co-ordination as it were. The race is not to the swift, nor the battle to the strong, but to those who can see it coming and jump aside. Like a frog evading a shillelagh in a midnight marsh.

Hunter S. Thompson, The Rum Diaries.

Chapter Three

Rehearsing a Rhapsody

Thou shalt claim for as long as is needed,
Nyaminyami for your River God

Life on Imire ran hard and fast and when things were too tense and we needed to shake loose, we would pack a bag of booze and climb Mutemwa Rock, a stone's throw from Imire on a neighbour's now unused farm. The brown flat, onion-skinned slabs of Mutemwa cracked underfoot as we raced to the top in a gruelling dash that left the smell of sweating bodies hanging sourly in the air. At the top we could forget the trouble Zimbabwe was in and look down on functioning dams, crops ready to be reaped, tractors and ploughs working hard in the fields, villages with little round huts and neatly thatched roofs and school playing fields; a jumble of children and colours. This was the functioning Zimbabwe and the one where bullying and coercion were not a way of life.

'We sure are living in some delicate bubble,' Sam, Reilly's sister, quipped.

'But it didn't used to be a bubble,' John said sadly, his brow furrowing and his sun-blocked white cricket lips straightening into a grim line. 'Zim was the bread-basket of Southern Africa. A country with promise and potential.'

I was the outsider who had become an insider. The outsider whose heart had been won over by Zimbabwe, its people, its beauty, and its fertility. Yes we lived in a bubble – I knew that as

surely as I was swallowing down cold beer that only served to rosy-up the view even more.

'Think positive guys,' I interjected into wandering thoughts, 'we'll get your old Zim back. At the moment it's not worth thinking about though. It's Imire that we need to concentrate on. Our elephant and our rhino.'

It was in this climate of mood, sometimes smoking hot and dangerous, sometimes melancholy and listless, sometimes energetic and conquering, that Rei and I built an infallible friendship that will span the years of our lives. Adventures were a way of life – life *was* an adventure and learning how to be a competent adventurer took practice!

But although life was hard it was always beautiful. Our days seemed wild and free and full of promise. In the evenings we would sit a-top majestic rock kopjes and dream; neither Reilly nor I could, or would dream small. The world then, lay unexplored at our feet.

These were the days when Zimbabwe's president Robert Mugabe was plunging the country into anarchy; and still is. People were arrested and tortured randomly; livestock and wildlife were in hungry demand as food was expensive. Inflation meant that neither we nor the indigenous people could keep up with prices. Queues for essentials were extraordinarily long and many people preferred to turn back to the old ways of hunting and gathering. Diamonds too were now huge part of the black market.

The volunteer program we had set up brought in some foreign exchange sure, but it meant that meals had to be provided and the constant struggle to feed the volunteers left us weary.

Fuel too was almost impossible to find and

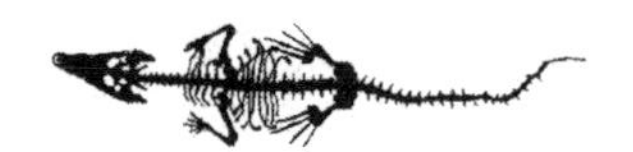

often we resorted to horseback riding when it was not available. Many neighbouring farms were getting rid of their horses due to the worsening conditions, so we had plenty of horses to choose from. We had a beautifully tended vegetable garden and a dam stocked with easily caught fish, which meant that we, at least, had a certain amount of bartering power. Trips to town would take three or four days and would involve the buying of as much of one item as we could get. We would then split this, be it sugar or salt, and barter it together with our fish and vegetables, until we had enough food or had run out of the ability to acquire more. It was on these shopping trips to Harare that we would meet up with Ace and tell him of all the grand plans we had dreamt up.

One day, beneath the fiery colours of a particularly awesome sunset, with the smell of rain on dust; Reilly and I sat on Castle Kopje and wanted out. Not forever; but we needed a break.

The constant erosion of the country, its people and its currency was affecting business; few volunteers were booking and there were constant fears for our safety. It was exhausting to feel you had to watch your back everywhere that you went. Poaching had become depressingly rife and the rhino murders preyed heavily on our minds.

'Let's take a break, do something for ourselves, learn some new skills,' Reilly said one evening.

'Something like what?' Ace asked.

Ace had travelled up from Harare. He often visited us out on Imire, and together we would climb the rock kopjes for sunsets and sunrises.

'Knowing you, it would have to be something crazy ... Kayak the Zambezi or something like that.'

Reilly paused mid sip and looked across at Ace. 'Now there's an idea. Imagine that, how long is

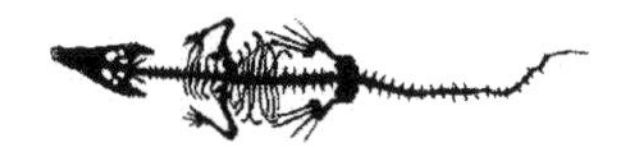

it? Three thousand kilometres?'

'I wasn't being serious, I was joking,' Ace looked a little worried.

'But seriously ...' Reilly looked at me.

'We could.' I paused and a few minutes of silence passed. 'I mean ... why not?'

In the beginning Ace thought Reilly and I were 'off our rockers' and perhaps we were. We had never kayaked before, had no idea how long such a trip would last nor how we would be able to feed ourselves throughout. We did not think of safety or backup and as for the money to fund such an expedition – well we hadn't seen real bank notes in over a year!

But the idea kept pulsating behind everything that we did, and finally Ace added an extra dimension. We would kayak the Zambezi to raise awareness and funds for the black rhino, and rhinos all over Africa. Slowly we began planning. We were all in: Reilly, Ace and I.

Raising money for our two-pronged cause began to fill every niche of our days. Reilly's tireless parents helped us to set-up two golfing weekends to raise money for the black rhino, a friend designed us a logo - **Row, Rhino, Row -** and better yet, at the London premiere of *There's a rhino in my house*, our expedition was mentioned (and likely promptly forgotten by those attending). We held a triathlon at the volunteer house on Imire and we talked. We talked to everyone we could think of. We figured word of mouth was our best bet.

We also started compiling lists of all that we might need on the trip, and then cutting down these lists again and again as we figured that we simply would not fit everything into the boats. With help, we also designed, printed and laminated a set of slides on rhinos and conservation that we could use in our talks to communities as we paddled

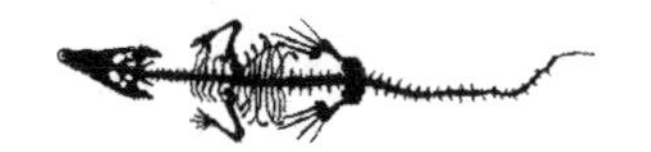

downriver.

The biggest run around was the visas. Where Ace and Reilly both had Zimbabwean passports, I had retained my British, as dual nationality was still not legal in Kenya. This caused no end of problems until; finally, as the date drew near we decided that I would 'wing it.' Ace and Reilly could get theirs stamped and I would do that later, after the expedition, by a friendly official that we knew on the Mozambique border.

The biggest problem posed to us now was Angola. As can be seen from the map, the Zambezi starts in Zambia and flows across the border into Angola, and back into Zambia before heading into the Caprivi Strip on the Zimbabwe/Botswana border, and then into Mozambique where it runs into the Indian Ocean at Chinde. Angola is not a country at peace and there was high risk of meeting rebels on the river. We decided that we would head up there and see what the word was once we reached that section.

''What do you know about the Zambezi Jamie?' Ace asked one day as we drove into town for another expedition related meeting.

'Not that much' I admitted.

'Perhaps we should take some time to learn something about it.'

Reilly and I agreed and after town we called by Norman and Gilly, Rei's grandparents, to tell them how our plans were coming along.

'Learn all you can,' Gilly cautioned us. 'Learn about the people of the Zambezi and understand how they survive. You will need these skills if you are to survive.'

'I agree,' Norman smiled, lightly tapping his pipe. 'Boys like you still haven't understood the importance of knowing all there is to know.'

Chapter Four

The Story of the Zambezi

Let the water under the heavens be placed together ... and let dry land appear

Norman was a great story teller and was rich in knowledge. He told us many a tale of the Zambezi. Where there were blanks, we filled them in with our own research, something both Norman and Gilly, adventurers in their own time, encouraged greatly.

Both the Tonga and the Lozi are people of the river. Lithe, velvet skinned fisher-people; they inhabit its upper and its middle reaches. In the sing-song language of the Tonga, the Zambezi is a 'great river,' while further downstream, in its middle reaches, it becomes the 'heart of all.'

Perhaps this mighty river took its name from a blend of these native tongues. Either way, the Zambezi is a lifeline that empties a basin spanning eight of southern Africa's countries. Its life-giving flow runs for over two thousand five hundred kilometres and its meanders tell of past floods and droughts, tribes and battles.

All down its length wildlife abounds, while the densely vegetated Barotse and Caprivi regions are natural sediment traps that provide hugely important wildlife refuges. Here, fish migrate and spawn, lechwe and sitatunga antelope dash amongst the shallows and hippo twitch their cat-like ears as they delight in the cool waters. Here and there other southern African rivers join the mighty flow in solidarity of its long journey to the Indian Ocean: the Kafue, the Luangwa, the Chobe and the Shire

amongst others.

In the wet season, storms as mighty as the river itself shake the black skies for miles around. Lightning bolts hang like skeleton trees in the sky, and animals take shelter wherever they can. Thunder hurled by the gods across the land, make the slate-grey baobabs shiver as the seed-banks prepare to push through the wet soils into a new, green world.

But the Zambezi does not belong only to the creatures that run wild along its banks. Man too plays a major part in the life of this ostentatious ribbon of life. Man and his machines, his industry and agriculture take vast amounts of water to feed his lifestyle. Massive hydroelectric dams, such as the Kariba and Cahora Bassa, provide the fuel that is needed for the ever growing populations. Issues over these dams still rankle dispossessed tribes. Now and again community disturbances and skirmishes bring to a forefront, the emotional and psychological upsets suffered by the tribes that still eke out a living in the surrounding regions. The dams too, act as silt traps, holding back the precious rich-alluviums that would, in years gone past, have reached communities living in the lower reaches of the river, which now tend to small subsistence crops in poor soils with little rainfall.

Victoria Falls, *Mosi-oa-Tunya*, or the smoke that thunders, marks the end of the upper reaches of the Zambezi with its mist-net of spray and precipitous figs that hang fearlessly over the edge and into the damp.

Below the falls, tourism has taken a firm hold, and suicidal tourists launch themselves on swings between the rocky walls of the gorge and off the 111m high bridge. 'Flights of Angels' take joy-riders up in Euro-copters to see the falls from above, while raft-loads of petrified punters run the angry

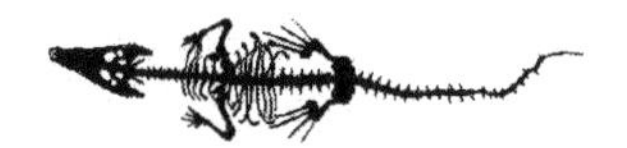

white-water of the constricted river. From here to the Cahora Bassa dam in Mozambique the river flows sometimes sedately, sometimes hurriedly, through its middle reaches. In this region, population centres are interspersed with tracts of wilderness where once the tsetse fly reined king. Both Kariba and Cahora Bassa support huge industries of kapenta fishing off iron-clad rigs that have a major impact on the economy of their regions. But in these middle reaches the legend of *Nyaminyami* lives on and populates many a story told by travellers in these diversely scenic lands.

Below Cahora Bassa the gorge contains the now-navigable rapids of Kebra Bassa that stopped Livingstone and his party from advancing up-river in the mid 1800's, and below these, the river meanders slowly to its mouth at the Indian Ocean. At the mouth, the grasslands were periodically drowned in flood years and so the grasses were protected from over-grazing, but now dam control of the waters has taken this natural cycle away and may account for the die-off of the coastal mangroves. Even now the coastal shrimp-fishing industry is in slow decline and this too can be attributed to the changing river ecologies brought about by Man in his quest for development.

The Zambezi still stands as a barrier to land based transport - it has never become a major trading artery as many an explorer had hoped - there are only a few vehicular bridges across its watery depths: the Katima Mulilo Bridge across the Caprivi, at Victoria Falls and Chirundu and in Mozambique, at Tete. Two car-ferry's cross at Kazungula on the Botswana/Zambia border, and at Caia in Mozambique.

In 1886 Frank Stanley Arnott, a missionary, was the first white man to discover and map the headwaters of the Zambezi. For the first time this

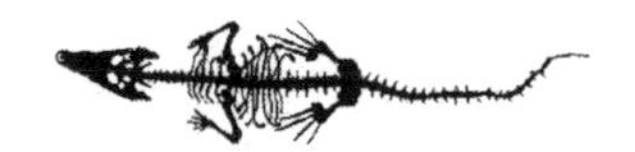

information was available to the public, and in 1906 Dr Walter Fisher journeyed out to the area and started a mission. It was here on this land, near this mission that our six month 'sojourn' began.

It was not that we looked at our trip with nonchalance in light of the dangers that we would face. We knew that potentially one of us would not come home. We had a good working knowledge of the bush and how to survive. Take Reilly for example: back when he was seven or eight years old his parents once dropped him and his brother Bruce off fifty or so kilometres from home.

'Right you two,' they said, 'out of the car. We'll see you at home soon. If you're not home by tomorrow we will come looking for you OK.'

These were the days before mobile phones and before parents sealed their children off from the outside world of adventure. Reilly and Bruce began walking through the farmlands; by seven p.m. they decided to find a place to stay but were chased off a property by dogs, and so twenty kilometres from Hwedza, they pushed on, made Hwedza, hitch-hiked home and were in bed by midnight. At school, Reilly was the kid with the lunch box full of grasshoppers and the kid who never seemed to do well in exams because he was looking at the nesting sun-bird outside the window.

(As I sit editing this book at Reilly and his wife Candice's house on Imire, there is indeed a jar of grasshoppers in the fridge!)

Ace on the other hand was more sensible. His DNA contained a lower percentage of Neanderthal. Ace always kept his head, never lost his temper and always came out on top. He was the 'city boy' amongst us – the Harare Hangdog, he knew how to get paperwork done, tick boxes and talk people round, he had learnt this from a family

that knew the secrets to success. But getting lost in a city – even a city like Harare (purple jacaranda lined boulevards) – requires skills. Ace knew how and when to get where he wanted to.

And me? Well I would say that I've had my fair share of comic but skill-teaching exploits. Back when I was thirteen years old I would load an old backpack with the spare car battery to which I would attach an old tractor headlight that I had converted into a spotlight. Hour after hour I would lug that heavy light and battery around on my own as I walked through the bush and across the plains looking for leopard, aardvark and jackal while pushing through or round herds of startled zebra and buffalo. And Mum and Dad? Well they were at home, warm by the fire waiting for my return – never past midnight they used to say.

Ace became the life pulse behind the idea; he was the motivator. Ace's father's company sponsored our kayaks and these were ordered and duly arrived from South Africa. Large unwieldy sea-kayaks that looked good to the eye and ignited the fire in us even more. The boats were long, sleek and double hulled which we figured would allow us to carry everything that we would require on such an expedition. A beginners' excitement took hold and we spent days in Ace's pool and on our dam at Imire trying to learn how to roll, but days into it we gave up and turned instead to the easier and more sedate use of the kayaks; fishing.

Several times over the months we took down a large map of the Zambezi and discussed the practicalities of border crossings, visas, un-navigable rapids, support and food. As the idea of such an expedition cemented itself in our heads, the practicalities seemed to grow smaller and smaller until we adopted the typical young African attitude

of 'we'll deal with it when we get there'. Parents despaired and yet encouraged us, really not knowing, like us, the sheer scale of such a journey. How would three twenty-three year old boys survive such an epic trip, they asked?

I rang my sister, Hollie, and asked her if she would cover for me at the volunteer project while I was away.

'How long do you reckon?' she asked.

'Couple of months, something like that, maybe longer,' I breathlessly replied. The excitement lay in just the idea behind the expedition.

She agreed, and Reilly too found a mate to cover him. We still had no starting date and no further practice in kayaking, but we had the naive confidence of all young adventurers, and the testosterone of all twenty-three year old boys.

Running the Zambezi was a side-effect of life as we lived it.

Chapter Five

The Journey to the Source

A River flowed out of Eden
To water the garden of Africa

Judy, Reilly's mum had just got back from the UK the day before, where she had spoken at the premiere of '*There's a rhino in my house*' and had mentioned *Row, Rhino, Row* with no idea of when our start date was planned.

'Hey Mum, you think you can drive us to the source this weekend' Reilly asked.

'This weekend? I'll have to check what's on, I -' she stopped suddenly and whipped around from where she stood at the kitchen table, Pog the warthog at her feet squealing for food.

'You mean the source? The Zam?'

'That's what I said.'

'I know, Rei, but wow, I mean do you think you're ready, have you packed? Are you fit enough? Have you told Candice?' Candice was Reilly's girlfriend and although not against us going, was pretty worried for our safety. And rightly so; any trip with Reilly, in those days, could often be termed a 'true cock-up!'

Only months earlier Reilly had ripped his ear off in a micro-lighting incident; he and his Dad John had rushed to a local clinic to get it stitched on again as they badly did not want Judy to see. At dinner, Reilly had carefully covered the ear with his sheepskin-like mop of hair and they had pretended nothing had happened.

'Reilly' Judy had asked 'anything of interest happen today?'

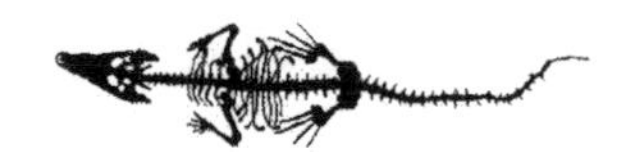

'Nothing interesting Mum, why do you ask?'

'It's just that your ear appears to have been stitched on backwards!'

Reilly looked theatrically aghast and Judy simply sighed and stitched Rei's ear on correctly with tools from her 'fix my children up' box!

The event had become the source of much hilarity as neither John nor Reilly, in their haste to cover up all evidence of their micro-lighting accident, had noticed the misplacing of Reilly's ear.

'Candice wants to come to the source too,' Reilly continued, 'she has arranged time off work. And Peeps and Stu Reid will drive their land cruiser so we can go in convoy. Ace's girlfriend Leanne is also coming.'

'Ah, so it's all sorted then? Asking me is just by-the-by is it, Rei?' I grinned at this family banter. A family I had been adopted into. Judy really was my Mum away from home.

'And Jamie, take that grin off your face. You two have to realise that this is a damned serious trip you are doing, not some knock-about-the-bush thing.'

'Yes Mum' Reilly said, and I was quick to follow.

'Yes Judy' but I simply could not take that grin off my face. This was it, we had a plan to drive to the source and we were setting off in less than seven days.

Four days later and it was a four a.m. start, on a fine June morning that got us to Harare in time for the morning black market, the norm in a very dysfunctional Zimbabwe. We managed to buy salt, eggs, bread, some meat, oil and tea with milk powder. We met up with the Reids and headed north up onto the Zambezi escarpment and down into the valley where the tsetse flies began their

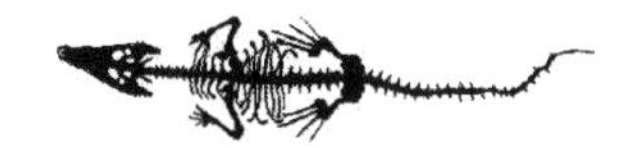

tireless biting. To the east lay the world famous Mana Pools National Park and Shitaki Springs and ahead of us, somewhere in northern Zambia, lay the source of the mighty Zambezi River.

Looking back on my diary now, I hone in on one entry that shows just how ill prepared we really were: *'I'm ready to go now and the short week we spent in Vic falls practising with our kayaks has, and will hopefully be enough for what we are about to embark on.'*

What we were about to embark on was a two thousand five hundred kilometre journey down a river which few had ever fully conquered, and here we were, three young men with a grand idea that we could succeed. We had heard stories, we had read accounts and we had made a point of speaking to anyone with river experience. In our minds, we were ready for the adventure.

The calmness of an early morning is wont to throw you back in time. Memories of other, cold, clear and purposeful mornings worm through your mind's eye and beg for remembrance. Years ago I recall waking early on the cold, misty slopes of the Aberdare Mountains in Kenya. We were on day seven of our somewhat bizarre expedition. We were on the trail of an animal that had evaded all for over a century (more about that later . . .)

In the faint light this search would be harder than finding a glass tear in the ocean. The Aberdares is a mountain range in Kenya where I spend much of my spare time, filling the days with fly fishing, waterfall gazing and exploring. This cold morning found me ensconced as close to the smoky fire as was possible with my good and now sadly passed companion, Dave Harries. We were peas in a pod, Dave and I, but in age we were thirty years apart. Dave was a legend of a man, full of the

experiences of life. Dave wore the sort of wrinkles that accompany every bush tale you will ever hear. He had a smattering of grey whiskers that told of hard work and sheer pleasure – after all, there is no time for shaving when the wilds of nature call! This time we were on the trail of the fabled Nandi Bear – a much maligned creature whose existence is tenuous at the very least and whose very being is "still adamantly unrecognised and un-described by science".

This creature was first seen in Kenya in 1912 by a Major Toulson, a military settler. He reported *"... One of my boys came into my room and said that a leopard was close to the kitchen. I rushed out at once and saw a strange beast making off: it appeared to have long hair behind and was rather low in front. I should say it stood about 18 in. to 20 in. at the shoulder; it appeared to be black, with a gait similar to that of a bear--a kind of shuffling walk ..."* In 1913 the district commissioner of Eldoret, a Mr. Corbett, reports that *"I was having lunch by a wooded stream, the Sirgoi River, just below Toulson's farm ... to my surprise I walked right into the beast. It was evidently drinking and was just below me, only a yard or so away ... it shambled across the stream into the bush ... [I] could not get a very good view, but am certain that it was a beast I have never seen before. Thick, reddish-brown hair, with a slight streak of white down the hindquarters,* rather long from hock to foot, rather bigger than a hyena with largish ears. I did not see the head properly; it did not seem to be a very heavily built animal." Later, in the building of the Magadi railway an engineer chanced upon some tracks that he sketched.

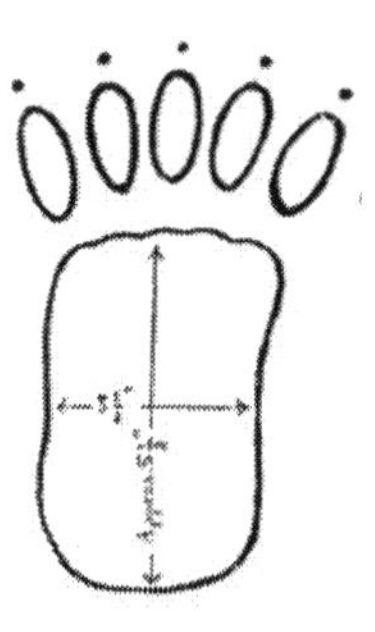

Footprint sketch of the 'strange beast' found by Schindler on the Magadi railway.

Perhaps they are those of the elusive Nandi Bear? I for one would like to think so. But the best reported sighting is that of Mr. Hickes who was an engineer on the East Africa railway in 1913. He saw the Nandi Bear on March the eighth of that year: "*It was almost on the line when I first saw it and at that time it had already seen me and was making off at a right angle to the line ... As I got closer to the animal I saw it was not a hyena. At first I saw it nearly broadside on: it then looked about as high as a lion. In colour it was tawny - about like a black-maned lion - with very shaggy long hair. It was short and thickset in the body, with high withers, and had a short neck and stumpy nose. It did not turn to look at me, but loped off--running with its forelegs and with both hind legs rising at the same time. As I got alongside it, it was about forty or fifty yards away and I noticed it was very broad across the rump, had very short ears, and had no tail that I could see. As its hind legs came out of the grass I noticed the legs were very shaggy right down to the feet, and that the feet seemed large...*"

Stories tell that the only captured, skinned and tagged specimen was shot within sight of the Nandi Hills forest by a man who collected specimens for the Natural History Museum of London. This fine

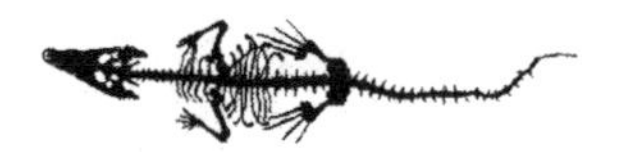

gentleman was on his way back from the bar, he had been called to the rescue of his wife who had locked herself in the house to escape the local drunk. This man stumbled across the beast and as his headlights momentarily stunned it he fired a lethal shot.

He was an efficient man and he skinned and documented all the details as well as snapping a few photos. All these were sent to Dr. Louis Leakey and from there sent to the Natural History Museum in London. By an unlucky chance the ship that carried the specimen to England sank within sight of the mouth of the River Thames. No evidence survived bar the shipping receipts and documentation, of which copies remained in Kenya. Since that day another specimen has yet to be collected. And now we, Dave and I were on its trail!

I was convinced (and still am!) that the Nandi Bear is in fact an as yet unidentified species of long-haired mountain forest hyena that perhaps used to roam the African forests two thousand years ago – a species that should not be disposed of quite so rapidly into the trash-cans of time. Perhaps Dave and I were the only crypto-zoologists in Kenya - no-one else knew the term obviously! (The study of animals supposedly extinct). Many claimed that the Bear was in fact the pied and bold honey badger, or even perhaps even a giant baboon. We however, had decided that; 1) if the Nandi Bear did indeed still live on, then its last refuge was the unexplored corners of the Aberdares and, 2) that some extra careful tracking was needed.

As we warmed our cold-bitten hands and pulled our scarves tighter around our necks, the air held the magic of anticipation, of adventure and of the unknown. The sighs that the dew-loaded branches made as they creaked in the frosty air were like sighs from our own hearts, and the creaks

of the old cedar trees were the creaking worries that we privately held.

Some months previously Dave and his son Jonah were driving high up in the mountains of the Aberdares and encountered a tawny brown creature that wore a shaggy coat and stood at five feet easily. Its face was that of a hyena but none that Dave or Jonah recognised. Its eyes stared back at them, dark as coal and the legend of the Nandi Bear resurfaced in their minds. The sighting alone brought an eeriness to the moment that made their hair stand on end. Dave stopped the car and momentarily the beast exhaled a warm cloud of air before it loped over the edge of a dip and disappeared into the thick bamboo forest. It was those ten seconds that had led us back here to search. If there was one place left where the Nandi Bear may live then it was up here in these ranges, where the cold and the bamboo played muted tunes all night long.

But back in Zimbabwe and the current moment meant that our spirits were high and the expectation of the forthcoming expedition was overwhelmingly heavy on our shoulders. As we came upon the Chirundu Bridge over the Zambezi - just before Zambia's border post - we pulled over to look. The Zambezi runs below the bridge in a deep chasm. This was our first view of the Zambezi since our fateful weekend at Victoria Falls and we hung over the railings as container trucks rumbled behind us making the bridge shake.

'Hey Jamie man, this looks like a pretty easy border post to get past' Ace noted.

I looked around and sure enough, the depth of the river meant that I could easily slip by while Reilly and Ace distracted the border officials and I could meet them round the corner a kilometre or so

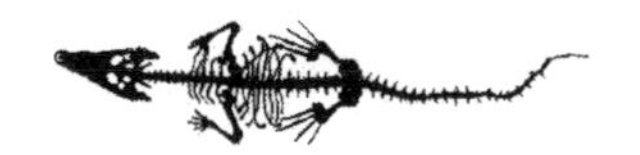

down the river.

'Yeah, you're right it looks pretty easy, Ace.' Having a British Passport in Zimbabwe at that time was simply not worth the effort. Old Robert Mugabe was blaming the British for all his problems.

Then we just stood in silence and gawked, trapped in a web of imagination. Little did I know that in a month when we did paddle below this bridge I would be 'bumped' by an aggravated hippo within view of where we were standing right now.

Time was precious on this long drive and we knew that the border crossing would take a good few hours, so we climbed back into the cars and rattled the hundred metres to the Zim border. Getting the 'out' stamps was easy, it was the 'in' stamps that we needed for Zambia that caused the problems. We had muddied up our kayaks to make them look well used, but Reilly had forgotten to remove the price tag on his, and the official noticed this quickly.

'You are going where?' he asked 'and for what are these long ... you call them boats?'

'We are going to visit a friend in Zambia who lives on a river. We are going to do some fishing and will use these kayaks so we can go faaar into the river and catch the big fish!' Reilly threw his hands wide apart and grinned, an attempt to appeal to most Zambians' love of fishing.

The official was having none of it. 'These kayaks are looking new, why do you need so many for fishing, maybe you want to sell them, no?'

We all assured him that no, they were not for sale and Ace and I moved forward to stand with Reilly.

'We are the ones sir, who will be in those kayaks,' I said.

He looked us over closely, dirty, torn shorts, uncombed hair and grubby t-shirts. That is when I

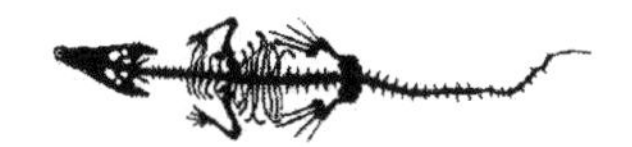

noticed that I was still wearing my 'Row, Rhino, Row' shirt and I hoped to goodness that the guy didn't notice. He stared for a long time and then wiped his palms together, straightened his neatly pressed collar and said:

'I will have to take this to my superior. You know here in Zambia we do things correctly and we cannot be letting you through our border without knowing for sure your activities here,' and with that he turned and walked into a closed office within.

When the superior arrived ten minutes later we could see we were in with a chance. The man was smiling and talkative and wanted to know all about our past fishing exploits and the fishing plans we had here in Zambia. We regaled him with tales from the nearby Mana Pools and told him that we had heard the Luangwa River in Zambia was renowned for its good fishing. He whole-heartedly agreed and sent us on our way with fully-stamped passports all round.

'Done and dusted,' Ace whistled.

It had taken four hours.

Just after the border post we managed to buy some more sugar, flour and bread and this did us for our first night's camping. The day finally ended after seven hundred kilometres at some self contained *bandas* on the Luangwa River. I look back on my diary now:

'That night was pretty, pretty special as we were on the road and on the way to the start of our dream.'

The Luangwa River is both world renowned for its wildlife and exceptional in its terrain, but we did not have time to hang around. I thought of my adventure with Dave and the Nandi Bear in Kenya as we rose early on the Luangwa to prepare for our drive to the source, but this morning was ever so

slightly different, I had never before felt as I did on this morning of leaving. In the early hours of the drive as the dawn broke, we all sat quietly with our thoughts and imagined the journey ahead. Here the anticipation outweighed all.

Day two was no less tiring and no shorter a drive. We again left before dawn and headed north through Lusaka to Kapiri Mposhi (where the Zambia-Tanzania train departs from) up to a farm in far north Zambia at a place called Kitwe. Here we planned to stay two or three nights so as to kill the road weariness that we had guessed would arise. This area of Zambia has been decimated by the by-products of mining: there are huge levels of deforestation that were sad to see and made us realise how lucky we were to have seen so many of the unspoilt parts of Africa.

I am not a 'people-person' and so avoid the built up areas, the peopled villages and the destruction that comes with population. My desires were to find and explore those places that were truly wild – the places where the wildlife had still not learnt to fear man and still stood, frozen, ears a-twitching when you came upon them, waiting for the signal to fight or flee. I think growing up; untamed in Africa, affords us the opportunity to see more places than the average city-dweller, whatever his income. Some places of beauty we may stumble upon en-route to another, while other places grow on us and give up their secrets ever so slowly and carefully.

The farm to which we aimed the convoy of vehicles was owned by a *rafiki*, a friend of Reilly's who had no belief that we would ever undertake this trip, and was absolutely dumbfounded when we turned up with two fully loaded land-cruisers and towing a trailer with three kayaks. It was a few days

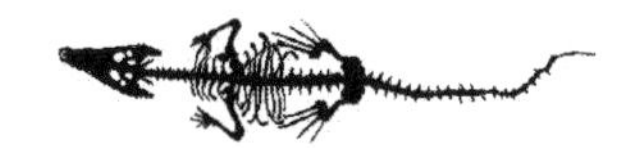

of pure relaxation and good food, something that we had not experienced in Zim for over a year.

'Jimbo,' Ace said thoughtfully one day as we lay swinging gently in our hammocks, 'you know we haven't even named our kayaks.'

'Hey, that's so true.' Reilly barged between us, knocking deliberately against us both. He flopped into his hammock.

'Well I'll call mine "*Ntwadumela*, he who breathes with fire".'

'Yeah you would,' Ace replied, 'and you'll probably set it on fire on the first night too! I'll call mine "*Yambezhi*".'

Ace's dig referred to when Reilly was five years old and had knocked over a candle in the thatched family home. The place had burnt to the ground that night, but since no lives were lost and possessions had once again been amassed, the joke was on Reilly – Only accident prone Reilly could have burnt his own house down!

'Yambezhi?' Rei asked.

'Yeah apparently it's the old Lunda name for the river and I've learnt one phrase in Lunda, "*shikenu mwani kunsulu ya Yambezhi*". Welcome to the source of the Zambezi' Ace spouted triumphantly, 'the heart of everything.'

'And you Jimbo?' Reilly asked.

'*Bahati*' I said. '*Bahati* because it means lucky in Swahili and it's the name of the place near where we live in Kenya – surely that will give me a guardian angel down the river?'

'Let's hope so' replied Ace. Reilly, being the avid naturalist that he is, was already caught up in his frog ID book.

On the last night our hosts conjured up an enormous barbeque and we ate like we had never eaten before. In our minds this might be the last good food we would get for a very long time and so

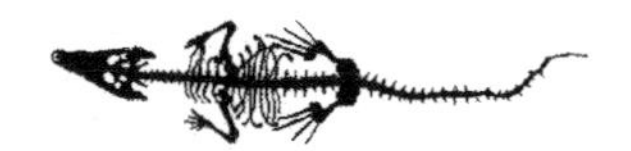

we made full use of it, and even stuffed our pockets with bits of bacon, *boerewors* and steak.

Would this be our last good meal? Would the mighty Zambezi be the end of us?

'Boys, why don't you stand up and tell us about your trip that you have planned,' one older guy, Colin Huddie, said as he gnawed on a chop. 'We want to hear this crazy plan of yours.'

He looked across at Zam, another old *bokkie* who sat twisting a chop in one hand and cleaning his beard with the other. He grinned.

'Yeah,' Zam said. 'Hold on, I'll get us all another whiskey and then you can start.'

And so we all stood up and between Ace, Reilly and I we stuttered through our plans to run the Zambezi source-to-sea and to live off the land the entire way down.

'No road assistance?' somebody asked, the glow of the firelight adding a glint to his eye.

'None' Ace said 'no time lines, no road assistance, no food drops, just us and a river and the bush – '

'– and of course a few joints and some beer,' Reilly added to much hilarity.

Some of the old *bokkies* that night at the bbq were experienced old *Rhodesian ridgebacks.* They had seen a lot in their day, grown up shooting crocs and hippos on their ancestors' farms, and many of them knew African rivers and their surrounding bush as we never had.

These old *Rhodesian ridgebacks* told us that night that we were going to die. They said that it was not physically possible to achieve this trip and, that even if we did someone would not return.

This brought it home to us, were we really attempting suicide here?

This was not new to us. Almost everyone we had spoken to thus far believed that we were

signing our own death warrants. At Vic falls, one of the old school veteran kayakers '- in fact he is so old school he is the headmaster -' people said, had told us that we were crazy. Mr Connolly had kayaked on almost every river in Africa and had seen more than his fair share of what Mother Africa can throw at you out there in her wilds.

'How you think' he had said, with a grin of a young boy, and the voice of a rasping Rhodesian, 'that you boys are just going to hop in a couple of kayaks just like that and paddle two thousand five hundred kilometres of a river, that even most of the best kayakers have never managed! And you have never even kayaked before - I just don't know – but I know you youth these days, you don't take no for an answer so I wish you luck, lots of it'.

We had long ago decided that if we listened to words like these, we would never complete our mission; we agreed not to let any words deter us from our goal. We took all that we were told with a good pinch of salt.

That night we lay in our hammocks under the starry sky, closer to our start than we had ever thought possible, and warm with several whiskeys in our bellies.

'Guys, you sure we are doing the right thing here? I mean we are only twenty three.' Ace asked.

'Yes of course' I replied, desperately not wanting anyone to lose spirit in the face of the stories and advice we had heard. 'Africa needs to be seen. Look at it man, it's being destroyed, too many people, too many mines. We need to see it before it's gone. Our children won't see this.'

Reilly added a drunken slur from across the way. 'Fuck awf, Jimbo, we are doing it because we are stupid and have some great idea that we will succeed. So it's decided, we are doing it. Anyway zip it, I'm trying to sleep.'

Ace grumbled quietly to himself and after a few minutes there was silence under the stars again and I lay there and silently wished for a safe trip.

It was good to get on the road again after two full days of negativity dressed up as joviality. There had been plenty of repairs to do on the two cars: punctures, bearings, belts and shocks. Now we were packed and ready for the last driving leg of the trip to the source. We said our goodbyes and headed out in two clouds of dust, North West to Pete Fisher's farm in the very corner of Zambia.

It was for this section that we needed a map. The going was tricky and the road nothing more than two wheel ruts at times. The *bitings* from the bbq were the sustenance for the trip. Before long our car, which was riding second, began, in true African style, to present problems. Finally after another two hour breakdown repair of slipped brake discs and with dusk fast coming in, we realised that the lead car had not stopped. We had been left behind.

Being left behind with mechanical problems was no new development and with several years behind us of being Rhino Charge entrants we were well versed in dealing with them.

The Rhino Charge is an extreme 4X4 event held in Kenya to raise money for the electrification of wildlife refuges and national parks. Initially, the electric strands kept in rhino and kept out poachers but now they keep out a ballooning population that is ever hungry for meat, firewood and money earning opportunities. Each year teams spend millions of brain-cells and argumentative words building and designing the 'perfect' extreme car. One that can be winched up cliffs, floated across rivers, carried on shoulders over boulder gardens and rolled without bringing the car to a mechanical halt, for

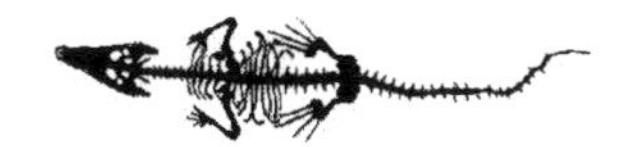

this is a competition that one wins by having the shortest mileage. Although there is still a time-barring it is the shortest distance between controls that must be found.

It was easy then to find a 'jua kali' (*hot sun*) way of fixing the slipped brake discs but how far ahead the farm still lay, we had no idea. We clanked through ruts and holes as we drove like rally cars, hour after hour in the moonlit night trying to catch up with the lead car.

Reilly began to worry that somewhere we had taken a wrong road. Suddenly, without warning there was a gate across the road and no turnings right or left. We skidded to a halt and Ace got out to read the faded clapboard sign fixed in front of the headlights; Hillwood Farm.

We were more than elated, we were buzzing. This was our destination and the source. We leapt from the car, weariness forgotten. Reading back on my diary I can almost re-capture the moment in my mind's eye: '*We jumped over the gate, we ran around the gate, we rolled in the sand, we opened the gate, we closed it again, we drove through the open gate, we drove back through it and re-enacted the moment for the cameras. It is one of the most exciting things to arrive at your destination with no real knowledge of where that actually is.*'

We could feel the clean, cold air, so we knew were close to forest, we followed the tracks of the last car that had been through the gate and arrived at the old farm house where Mr Reid (Ace's Dad) was worriedly sitting waiting for us.

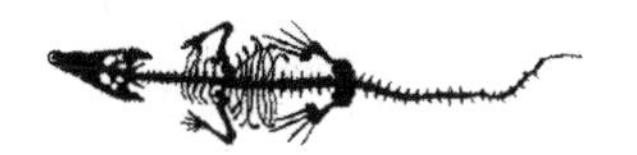

The house had belonged to old Mr Fisher's mother who had died. He had cleared it out and the house now lay empty on one of the most scenically stunning farms in all of Africa. Fisher was a descendant of Frank Arnott, the early explorer. His family had lived on the farm for so long that when the colonial masters were drawing up the maps and boundaries of the countries within Africa, he was asked if he wanted to be part of British Northern Rhodesia (now Zambia) or Portuguese East Angola. He decided that he had a closer affiliation with the British and so, as per his request, the Zambian border encompasses his farm; his farm boundaries now form the Zambian-Angolan-Congo border.

We hung our hammocks in the garden - it would be outdoor sleeping from now on we vowed, while the folks and girlfriends slept in the house on their ground mats. Mr Fisher was away with hunting clients and would be with us the following evening.

Although weary, our body clocks were alerted now to the sun, sunrise broke and *'there in front of us was the Congolese rain-forest; we had actually reached the forest. It was a huge set of glades where we could see the stream that was to become the big Zambezi River. At that point the rain-forest started and there, behind the Kaleni Hills lay the Congo, in my mind the most difficult river in all of Africa. What we were attempting was tame.'* We could see hundreds of kilometres of forest. We woke up that morning with smiles on our faces; for sure we were seeing stuff that we had never seen before.

There were a lot of giant sable on the farm and these were a first for us. We spent the morning

walking. We walked down to the stream that was the Zambezi and we leapt in it and swam in the cold, clear waters. We spent hours just staring out at the beauty that lay before us.

Giant Sable is a rare sub-species of the sable antelope that is endemic to Angola. It is the regal emblem of Angola itself and is held in high regard by the people of the region. As with many creatures that so symbolise life and beauty, the giant sable is now critically endangered. Its habitat of tropical, high altitude forests with their rich variety of leaves and tree sprouts are fast being cut down and populations moving in quick. It was believed that few had survived civil war in Angola, but in 2004 a series of camera traps documented the last remaining herds. These noble creatures are free to roam across this beautiful farm in Northern Zambia, one of their last remaining refuges.

Eventually we set off for our destination which was to be, in effect, our start. On the way up to the source, we straddled the border, right hand wheels of the car in the Congo, the left in Zambia. After coming so far we expected something wild and remote but the source was a national monument with car park and a dilapidated visitors centre complete with curios. The infant Zambezi simmered a short walk away, at the base of some wooden steps; there were none of the big trees and no massive *ficus* buttress roots that we had imagined either. All around us were thin, almost puny, tree trunks; anorexic giants, which raced up to the canopy. Beneath the clear water and yellow-and-black orb spiders, the thread-like fibrous roots seemed to push and jostle the leaf litter as they entwined. There were one or two bigger trees about and lots of brown umbrella fungus. In my mind it was hard to register: *this* is the source. In some

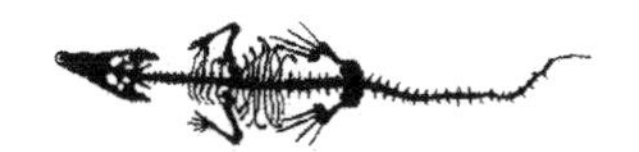

ways it was almost an anti-climax and I had to fight against the feeling of seeing another spot in Africa that was falling victim to decay. But as we sat quietly, at first on two horizontally growing tree-trunks, the original source tree that had pushed branches out to grow as thick as the trunk itself in their quest for sunlight, a young Zambian man with bright white teeth and dimples beneath his wide eyes sang songs that he accompanied on his guitar. After a while we gave him a few *kwachas* and he musically disappeared. Now the magic began to clamber into my heart.

'Hey, Reilly man, look I've stopped the flow of the whole of the Zambezi.' Sure enough my cupped hands had dammed the flow of the tiny trickle of cold, clear water. Judy, who is always one for magical moments and little ceremonies accompanied by tears of joy, decided that we must each fill a bottle with water from the source, tie this to our kayaks and pour the water into the Indian Ocean when we reached it, a symbolism that she believed would add to our safety. We agreed whole-heartedly. Miraculously an old *m'ganga* (witchdoctor) arrived with his lotion and potion bottles to collect water from this sacred spot.

'Hello *baas,*' Ace greeted him using the formal term of respect.

Judy's face lit up and she pulled the poor old man aside and whispered hurriedly to him. A great toothless smile lit his wrinkled face and he nodded emphatically.

What ensued was a very solemn consecration ceremony in which the *m'ganga* blessed our journey and wished us luck on our travels to the sea, a place that the *m'ganga* had never seen and could not imagine despite our colourful descriptions. When he had departed we mulled over some critical questions amongst us.

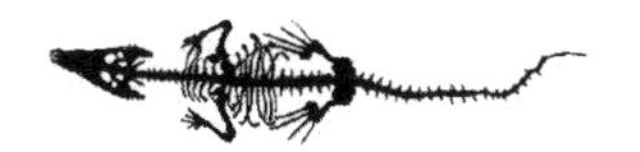

'How long do you think it will take us?' Reilly asked the uppermost question in all of our minds.

'Maybe three months,' Ace replied.

'Hmmmm, what do you reckon is going to happen?'

No-one cared to answer.

We then walked that section of river, down to where it opens up slightly and then walked back up to the old farmhouse where we were to stay another night.

All through history man has desired to find the source – not only of rivers but of anything that he meets upon his path. And yet to me, man is the source of all destruction. The source of the Zambezi was of course known by the native peoples for as long as they had been a part of the intricately woven parcel of Africa, but in told tales of discovery it is only 'explorers' who can claim written rights to discovering. The explorer who was lucky enough to put his name to this source was Frank Stanley Arnott in 1886. He names the marshlands between the basins of the Zambezi and Congo rivers the 'sloppy country.'

And so here we were, entering into the 'sloppy country!'

As we reached the farmhouse we found that 'Old Mr Fisher' had arrived. He was an enormous man, a reflection of the enormous continent of Africa. He had seen hard work and good rewards in his time and loved his life out in this God-forsaken corner of Zambia. The news he gave us dropped like stones into our imaginations. In my diary I wrote; *'Old Mr fisher came round and called a meeting of us all. He said that five days previously he had been called to help retrieve the body of a twenty something year old British boy who had been in Angola looking for*

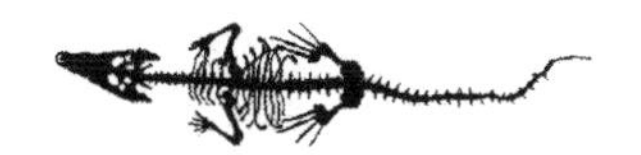

diamonds. Somehow he had ended up being shot by rebels and as the British government had not wanted to go into Angola, the family had rung Pete and asked him to go, which he duly did. He said that he had no desire to go and rescue anyone else and with the rebel activity currently going on in Angola we were not to do that section of the river.'

Now when this is said in front of mothers, fathers and girlfriends, things don't go down so well even if you mix it slightly with whiskey and the last of the good cheese.

Afterwards Reilly, Ace and I sat out in the garden and argued about whether we should ignore the advice or not. We knew that we had to respect the wishes of all involved but not doing this section meant cutting out two hundred and forty kilometres of river – so how could we claim source-to-sea?

Judy came to join us and pointed out the full moon. In the silence that fell she asked if it was really the source-to-sea claim that we wanted or the adventure of river travel. 'You boys are not about labels and accomplishments, you boys are about experiences and fire-side stories.' Quietly she took off all her silver jewellery and laid it in the moonlight on the grass. Her whisperings continued.

'Ultimately boys, the decision rests with you. But don't forget that your lives are shared and that the people that share them need consideration too.' She asked us to look at the moon and think of our safety and the safety of our would-be rescuers.

In the end we realised that through their sponsorship as Northern Haulage, Ace's parents' company had put in the majority of funding for the expedition. Without them we would not have been able to get even this far. Ace's parents did not want us to do the Angola section, and as we were a team,

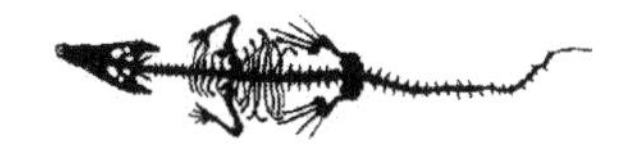

we made the final decision to drive around this section and launch our kayaks at the point that the river flowed back into Zambia. We agreed that we would come back sometime in the future and kayak the section that we had now decided to leave out. It was an unfortunate setback to our expedition, but we felt the decision was wise in the light of the complicated mess up that was playing out in Angola.

As I slept that night I thought of Angola and the diamonds that were the root of so many of its issues. I thought of its people who lived in such abject poverty although their country was rich in minerals and gems and I thought of how they were exploited by those few whose ethics and morals were limited.

Angola and her minerals, Zimbabwe and her land-grabbing, Zambia and her deforestation, Mozambique and her civil unrest - *'earth has enough to sustain every man's need but not every man's greed,'* I whispered to myself as the silver moonlight haloed all the beauty that lay around me.

I remembered a time that diamonds has reared their head in my life too. In 2007 I had booked a direct ticket with Kenya Airways home to Nairobi in Kenya. In Harare I was one of only three passengers waiting to board the flight in an airport that was crumbling just like the country all around it. Air traffic was restricted to daylight hours as the many power cuts that plagued the country meant that the runway lights were not often illuminated.

We took off into the clear sky and I watched with interest the changing landscapes below. After about two hours flying we appeared to be descending. I looked around me but there were no air hosts present (there were no hostesses). Soon we came low over forest and circled a broken tarmac runway. We bounced and skidded as we taxied to a

standstill and still there appeared to be no-one about to question. The doors were opened and a small staircase wheeled to its opening. A voice came over the intercom:

'This is your captain speaking. Please remain in your seats; we will be on our way again very soon.' And that was it. Nothing more.

Around the plane was now a hive of activity; military jeeps, men in police and army uniforms and many well-dressed, bling-sporting Indians. Tens of small wooden boxes were loaded into the holds below, while the seats within the aircraft filled up with Indians. I attempted to question my neighbour as to where we were, which country? What were the small boxes being loaded and who was he? The man declined to answer even a single question.

Soon we again took off and on schedule, landed in Nairobi. I left the plane expecting long queues in the immigration/arrivals hall but instead was greeted with silence. Only me.

Only while I was standing at the baggage carousel musing did I realise that many of those Indians had been wearing diamonds – diamond encrusted wrist watches, necklace pendants and small diamond studs. To this day I can only conclude that we landed, perhaps unknown to the airline carrier, in either Angola or the Congo to pick up a shipment of diamonds that was bound to some country in Europe or the UAE.

As I lay swinging in my hammock in the Zambian night air, I dreamt too, of the small stream that rubs three sides of our garden at home in Kenya, I dreamt of our futile search for some supposedly age old cave-art that lay hidden somewhere in the steep cliffs that backed that farm. Even camping in your own garden can be magic and I remembered twinkling fireflies and the hooting of an owl I had not heard before. Judy, as always, was

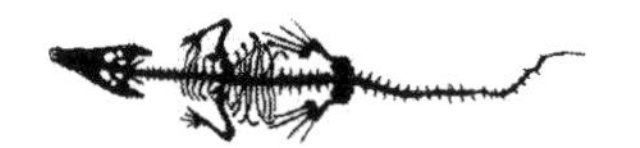

right. We were not about source-to-seas, first-descents or claims to fame of any kind. We were about journeys, adventures and wildlife. The river would bring us these three in vast quantities.

Dreams and diamonds aside, I woke early the next day and we waded the short, shallow section from the source to the Angolan border and then loaded the cars to head to the next put in, Chuvuma Falls. It was a horrific journey beset with punctures, ceased wheel bearings, and at one point we came across what appeared to be mad pygmies in the road. They did not look as if they would let us pass. As we got closer they appeared to be dancing in some form of ritual, and it was not until Reilly and I joined them and danced with them for about an hour that they let us pass, bemused and exhausted.

That night we stayed with a missionary, Bob Young, who had been in the area for over fifty years. When he heard of our plans, he whispered that he had watched one of his five sons be taken by a crocodile in the pool section at the base of Chuvuma falls. Right where we planned to put in. He advised us not to put in there at all. In fact he advised us to call off the trip. That night was not restful as all of us stayed awake wondering about the future.

But the show had to go on, and the next morning we drove as close to the pool at the base of the falls as we could, and unloaded everything from the cars. We assembled the kit in front of a growing horde of locals who wondered what the hell we were doing. When each kayak was finally packed, it weighed eighty kilograms without us in it!

'Can we pay some porters to help?' Rei asked the assembled crowd.

Smiling men scrambled to the front with whoops and cries - 'me! Me! Me!' they cried, wanting

to be a part of this interesting happening.

Ace selected six muscled men who couldn't wait to load the kayaks up on their shoulders. The excitement was such that first one man then the other scrambled to lift one end of the heavy kayaks onto their shoulder only to have it slip back to the ground as the bystanders laughed shyly into cupped hands. We all grinned and organised the porters so as on the count of three each man lifted his end of the kayak.

'One - two - Three,' and loaded up we began to walk.

CRASH!

All three of us whipped round to hoots of laughter and two sheepish porters! The two men had been facing in opposite directions and without thinking both had walked forwards and out from under the kayak! The boat was dropped heavily and in true African fashion this somehow did not end in us needing a new kayak before we had even begun!

Once we had made sure each porter was facing the same way and that each were headed in the same direction, our boats were finally deposited safely on the shore of the pool below Chavuma Falls.

Each one of us peered over the outcrop-like shelf of rock that made Chavuma Falls. The roar of the water meant no-one could talk. I saw Ace mouthing his wonder, eyes wide and mind alert. *Aye yai yai yai ya,* he mouthed. Was our expedition going to come to a grinding halt before we had even started? We looked across at each other and shook our heads. No-one in their right-minds would run these falls in a sea-kayak. We took solace in that. But the fact that we had even contemplated running it was a sign of things ahead, of things we were to attempt that we should have left undone.

Just to confirm our choice Ace sent a log down the rapid. It got caught in some pull-back

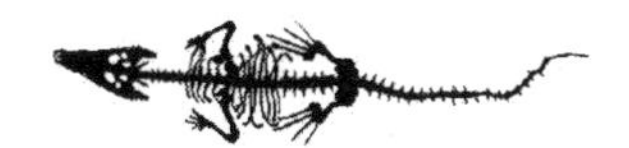

current beneath the falls and we never saw it again. Ace shouted across the roar;

'On wards and down-wards boys,' a grin lit his already sun-bronzed face.

'*Chav-ummm-haaa*!' the falls roared back at him as the water was forced through the narrow rock openings.

With a farewell bible each from Bob Young and hugs and best wishes from our families, we cracked a bottle of champagne across the fronts of our kayaks, officially named them and launched midst much ululating and shouting from the locals, and after Candice had hurriedly pulled Reilly aside.

'Rei' she said 'if there is one thing from this trip that would be special to me, it would be that bible. Keep it safe and it will keep you safe. Then will you give it to me at the ocean?' Reilly gave us the timeless look that spoke the language of all men - *women.* Reilly never said no to Candice.

We were finally on our way. The cars hooted until we could no longer hear them and that was it, we were on our own.

Part Three

Water notes from the upper reaches

'The principal difference between an adventurer and a suicide is that the adventurer leaves himself a margin of escape (the narrower the margin the greater the adventure), a margin whose width and length may be determined by unknown factors but whose successful navigation is determined by the measure of the adventurers nerve and wits. It is always exhilarating to live by one's nerves and toward the summit of one's wits.'

Tom Robbins. Another Roadside Attraction

The best journeys answer questions that in the beginning you didn't even think to ask.

Chris Malloy 180° South.

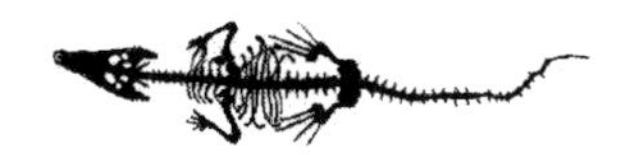

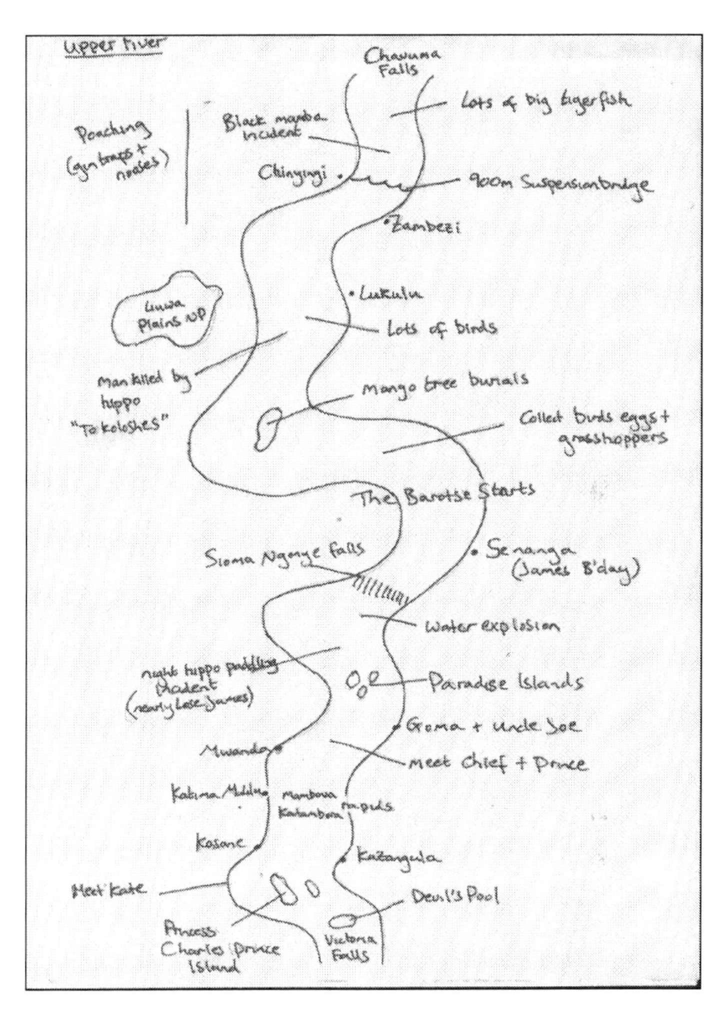

Sketch Map of the Upper Zambezi

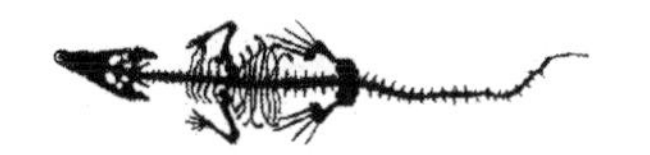

Chapter Six

Chavuma Falls – Barotse

Thou shalt kill only for food

And so the river journey had really now begun. It was the middle of June, our minds clear and our strokes strong and confident as we paddled with the current and around a bend in the river, leaving the thundering Chavuma Falls behind us. Ahead of us snaked more than two thousand kilometres of river still to paddle.

The feeling was overwhelming.

There would be large rapids along our way that would test our as-yet-untested paddling skills.

The long, slow stretches of calm water would test our mental tenacity.

But all of the Zambezi would provide adventure.

The journey ahead coaxed and beckoned us both at night as we sat around the camp fire, and during the day as we fought spandangled tiger-fish on our lines. River-side trees whispered down-river secrets in our ears, and river currents curled stories around our boats and paddles.

Those first days were tough on our bodies as we adjusted to the regime. Muscle built up where it was needed, and little by little we became accustomed to the hard monotonous slog that is flat water paddling. Our stomachs too, had to adjust to the food levels as we increased the amount of carbohydrates and protein rich fish and decreased the fresh fruit and vegetables by sheer necessity of availability.

The winter grasses that fringed the river

were burnt golden and were both woody and spiked. They pushed through ivory sand that was cool and hard to the touch. Those first few days on the river were magic, we didn't have to hunt and gather as we had enough stock to last a week. Soon though, it became pitifully obvious that we would need to work hard for our daily ration of *nyama* or meat.

Sometimes when my stomach grumbled I was thrown back to Bush-Academy days in South Africa when we were left out on our 'survival' weeks.

With Ray Mears and Bear Grylls now 'live' on prime time TV, most Western children believe they could 'give it a go and survive' but it's a lot harder than it looks on TV! There are places in the Congo where the pygmies live only to their mid-thirties due to the sheer energy levels that are required for hunting and gathering, as well as the danger involved. Many die as they plummet to the ground from the gallery trees where they collect honey.

It was 2007. We were green about the ears and Bush Academy in South Africa was pushing on our comfort zones.

'Survival Course,' our teacher Fritz Breytenbac had gruffly addressed us, 'is no joke.'

We sat, a group of boys and girls, thinking we *were* already survivors.

'Some of you are going to succeed out there and some of you are going to fail. The rules are as follows: Any game is fair game within the park boundaries. No rhino or elephant. Rare and endangered species are off the menu! Vegetarians will be given one pumpkin. No matches or lighters, one knife per person and only the clothes that you set off in. Wander where you like, how you like, stay safe, go well and good luck. Understood people?'

We had all mutely nodded, apprehensive at what was in store for us.

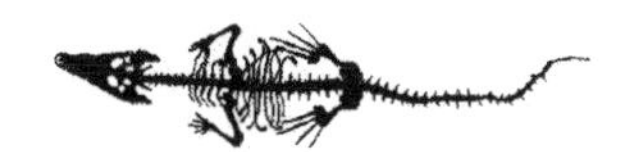

Off we set, a group of five students with Lovat from Kenya, Reilly and I. We conferred hurriedly as the starting shot was fired and decided to head north to the flat plains where ostrich were often seen.

'Warthog; hmm they're too dangerous,' Reilly mused, 'antelope; too fast and buffalo too big and dangerous.'

'Our best bet is a small ostrich, most of the chicks have hatched now, they must be about a metre tall by now,' Lovat interjected.

'How are we going to avoid the mother?' I asked.

'Not sure but we'll manage somehow, ostrich don't have the biggest brains. Let's go!'

And into the afternoon we had marched, brazen and cunning.

The afternoon had turned to evening and the evening to night and still we had seen no ostrich. Night came upon us and while we managed to get the fire going there was no food to cook.

Up at 'sparrow's fart,' we marched upon a coppice of thorn scrub that encircled a clear area that could potentially be our ostrich trap. We squatted on our haunches and scratched out in the dirt a plan of action, all hinging on there being any ostrich about at all!

As we conferred, an ostrich mother and her seven chicks appeared from behind a knoll; Superb!

Although they were over a kilometre away we could see that the chicks were nearly sub-adult, about six months old.

'Alright,' Lovat called, 'get into your positions!' And so the ambush was laid and all took their places. The wind blew in our favour, towards us and head on, brushing the golden grasses that grew tall with its touch. Quietly we cut some saplings into *rungus* or throwing sticks that are

bulged at one end. We could see the ostrich slowly moving towards us, the chicks chirping as they scuttled after grasshoppers and flying ants. Reilly, Lovat and I were nominated as the throwers and we concealed ourselves in the waving grasses at the opening to the glade, creating a funnel with our positions. The other two stealthily circled the ostrich carefully from behind.

There was silence among us as we all waited. Chirping, the ostrich chicks scampered unwittingly towards us. Tension as the wind blew towards us.

Then the wind picked up another call. A '*wheeeeeee*.' The high pitched call of a bird of prey.

Suddenly and without warning the ostrich family panicked and, necks raised high, they ran at us at high speed. Our careful planning botched, we broke cover - the ostrich caught sight of us and ducked as they spun to run away from us.

We had no choice but to attack.

I hurled my *rungu* and missed. Frantically I tried to circle round to collect it from where it had fallen.

'Watch out!' Lovat called as he threw his. As Lovat called, the other two members of our team burst on the melee panting and with extra throw sticks.

Panicked birds, feathers, dust and torn up grass were twisted together in shouts and calls. Runners tripped and sprawled. Birds leapt up in frantic dance steps as some spinning throw-sticks hit their mark.

The smell that is so noticeable when predator kills prey was in the air. A sweet musk that spoke of fear and a fight for survival. In these moments of absolute action it is hard to describe exactly what happens. As the ostrich fled we realised that three of the young had succumbed to our blows. Too many for us to eat.

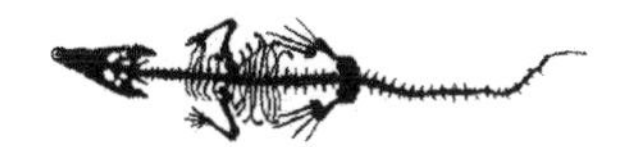

Four of us turned to the task of plucking and preparing our birds while one ran to call some of the others with whom we could now perhaps trade or share our bounty. Our runner came back in a hurry.

'Come! Come look what I've found.'

We rushed after her thinking some predator had smelt or heard the battle and was cashing in. Suddenly she stopped and crept closer to a bush which she then peered around fearfully.

Our eyes followed her pointing finger and came to rest on the last crumbling structure of what must have been an old abandoned ranger post.

'Honey roasted ostrich with wild ginger,' she quipped, 'imagine that.'

Our bellies rumbled and our mouths salivated with the thought of it. We collected fistfuls of green grass and lit them with the help of flints that we had fashioned from stone. We thrust the smoking torches into a crevice below the old tin bath that sat upon and within a still firm concrete structure. The bees succumbed to the smoke in sleepy short flights and it became easier to avoid their painful stings.

Soon it was safe enough to crack into the hive, a task that proved to be very difficult. It took three heavy blows with a boulder that four of us held to batter our way into the cement. As the cement crumbled away under pressure a vast hive was revealed, that must have been in existence for over ten years – at least! The bees inside were feral – they flew wildly about in a rage that culminated in red hot burning stings across every inch of our bodies.

We laid more green brush on the already smoking fire to create a haze in which we found breathing difficult, our eyes stung and tears left smoking trials across our cheeks – but nothing stopped us. We reached out into the hive grabbing fistfuls of honey-filled comb in swollen hands. Soon

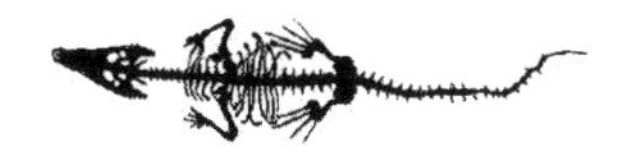

the palm leaf bag that we had fashioned was full and we abandoned the crime scene at a run, trailed by angry bees that whined and buzzed their anger at us, their predators. We had opened the way for the honey badger, the pangolin and the porcupine.

That night we sat beside a spitting, glowing fire, full and content, all groups united with a bounty that made a meal fit for kings and queens, a tasty spread that would last the entire five days of Survival Course.

As the stars moved across the sky the five of us sleepily scratched swelling bee stings and mumbled through swollen eyes and lips, shifting our puffy hands and feet to get comfortable.

Back on the river and we had absolutely zilch intention of turning vegetarian on this Zambezi expedition. We were game for anything, you name it, we were planning to eat it: Fish, crocodile, birds, monitor lizards, snakes. We each had a catapult and we shared a spear-gun. These two items, together with our Leathermans and packs of fish-hooks, were essential to surviving the trip.

Winter was still in evidence despite the warm, clear days. The nights were chilly and the water cold. We all suffered from sunburn badly in that first week until we got the full measure of the sun and learnt how often our sun-cream needed to be reapplied. I fashioned an 'eagle's beak' on the nose-piece of my sunglasses so as to stop my nose from peeling from under my bush hat, and both Ace and Reilly wore wide brimmed hats, Rei's of straw and Ace, who was an avid cricketer, a hat of his white, Sunday best.

Sunburn, sunstroke and dehydration may seem like light-hearted malaises to sun-craving travellers but on *any* expedition they are dangerous and at times even lethal. *"Fever, sunstroke, and fatigue," says the missionary explorer Dr. Krapf*

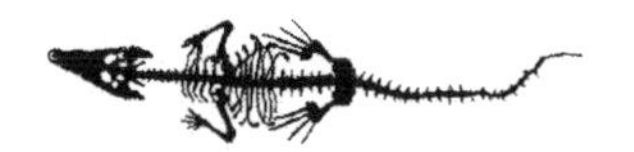

"nearly killed me; and I quite expected to have found a grave in the Nubian Desert." Such were the dangers of dehydration.

The river banks were lined with *Combretum erethrophylum*, the river bush-willow. The branches hung low and wide over the banks and made for a picturesque paddling vista. Paddling underneath these swaying fronds gave much needed shade against the glaring sun that we had not yet acclimatised to. The bush-willows were full of bird life and as we drifted lazily downstream, we tested each other on the names and calls of the birds around us. Doves, always in pairs, cooed at us as we passed by, cranes called at us from above, 'Further down! Further down!' White flashes of egret melded with the constant movement of cormorants, and the little bell-like nests of the weaver birds were tended to by vivid flutterings of yellow.

The river was a conveyor belt of movement and almost invisible locomotion.

It seemed an age ago that we had set out on this adventure and yet these were still our first days on the river. Camping on those first few nights seemed almost vast, a word that I'm not sure is vast enough in its description. It seemed like it was just ourselves and the river. The pockets of settlement were isolated, which gave the welcome feeling of space and freedom. There were plenty of trees that cast dappled shade, for stringing up the hammocks and we established a rhythm of life that we would sustain throughout the trip.

Sadly though, the surrounding wildlife had been poached out by the resident natives for the local bush-meat trade. Huge fields had been cleared for subsistence crops and majestic giants hewn down to make *mokoros* (the traditional canoes). We found several gin traps - toothed and rusted half circles of killer iron that snapped grimly shut at the

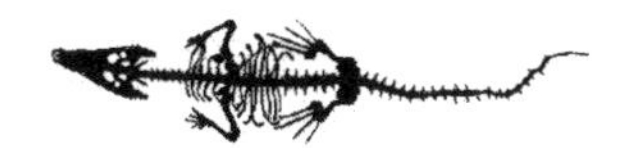

slightest pressure. We saw loads of fishing nets and lines and at night, shots jarred the darkness. Often we would come upon a hunting party with six, seven, eight dogs on the riverbank lugging bushbuck, duikers and steinbuck strung on poles between them.

Subsistence farming looks to nothing in the future except survival. Where populations have expanded beyond the land's carrying capacity, rape and pillage of the natural resources will occur as a matter of course. All over Africa, very few governments have taken steps to provide other forms of income, and now exploding populations have no choice but to fend for themselves. What we saw was the direct result of this: cleared lands that became more infertile with each passing year, and wildlife populations that were teetering on the brink of local extinction. Rivers were altered, their courses perverted from their ancient narrative, and age old sites of worship or tradition were lost in the name of 'progress.'

Progress was something we discussed often in those first few days but answers were few and far between.

One happy paddling day Ace and I managed to get a good few hours ahead of Rei and we pulled up to wait for him beneath some dappled trees.

'Hey Jimbo,' Ace called across to me from his hammock as he swung to and fro in the gusty wind, 'we got *bhangi* and we got some tobacco, but these rolling papers are not going to last a week.'

'Don't worry Ace I've got a plan,' I looked across at him from my own hammock and winked.

'Oh yeah, what's that?'

'We'll use the pages from the bible, I already tried it, they are perfect. And we have a bible each so that should do us all the way to the Indian Ocean!'

'*Lekker Bru*, that's *hundreds*' Ace grinned widely.

In hindsight it was unfortunate that Rei had missed this important conversation . . .

One of our favourite games in those early stages was choosing an accent and sticking with it for the whole day. Later, it became a competition to tell a story in our new accent, and each night our camp spat out guffaws of laughter like embers from a fire.

En-route we stopped to explore a mission in Chitikoloki where Americans, Englishmen and a Scotsman worked practising plastic surgery, dentistry and treating leprosy. The mission had been running since 1914 and its sheer efficiency astounded us.

Our efficiency could not match theirs - our meat ran out on day four, except for the biltong which was highly prized and strictly rationed for our own sanity. Biltong, or salted sun-dried meat gave us back the salt that our bodies craved after a long hot day of sweaty paddling.

We were pretty fired up to catch our first meal, and we watched carefully for any potential meal movement, but our few catapult shots had missed bringing in a pigeon or two for the pot.

On about the seventh day, Ace had paddled ahead and Reilly was trailing behind in a dream when I saw a long, black coiled shape up on a limb on one of the *Combretum* trees. I did a few back strokes to pull myself closer and my suspicions were confirmed.

'Reilly. Look. A mamba. A black mamba!' Reilly pulled in behind me and we peered up into the bush willow, looking at the snake.

The black mamba is the most feared snake in all of Africa. It is fast; it is fearless and it is lethal.

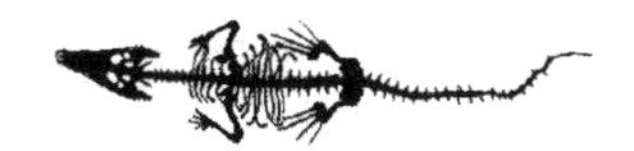

The coffin shaped head is a sign of what is in store for anyone who takes a mamba lightly. Even the smallest amount of venom is enough to kill, and out here, a trip to hospital was not even conceivable.

But we were unbeatable die-hards who could live forever, like young men all over the world. We debated quite a few things at this moment including its length, which turned out to be a good ten feet, and how well a mamba can swim, which turned out to be pretty well. Then the question arose:

'Reilly, should we kill this thing and eat it?' I could see the same thought had surfaced in Reilly's mind.

Reilly was unsure of how to go about killing it. We had both heard plenty of stories about how tasty a mamba is and had eaten it once before at Bush Academy in South Africa. We had been driving cross-country when the car in front had run over a mamba. We had taken it back to *digs (student quarters)* and after marinating it, had fried it up and served it on skewers! This memory had been trigged in both of us and we now unanimously decided that this tasty morsel of snake was officially on the breakfast menu.

But how to kill the mamba while still maintaining a tight hold on our own lives?

So we schemed, and we schemed, and we hatched a plan.

We pulled into the bank and collected smooth pebbles for out catapults. In my mind I always think of a black mamba as a racing car with its V8 engine idling. As it warms in the sun the engine reaches optimum temperature and the mamba can move from zero to sixteen kilometres per hour in less than five seconds. This lethal snake was not to be messed with.

And here we were, about to mess!

Whatever we did, we were aware that we

had our lives to lose.

Aiming and shooting accurately while sitting on a moving current is not an easy task, and so it was with many misfires that we began shooting at this mamba. We would back paddle and then, as the current slowly brought us downriver, we would wait until we were level with the mamba's head before firing.

There was no way we could simply wedge ourselves into the bank and shoot, as then we were directly under the mamba, and that would be increasing our chances of death by a huge margin, and so down we floated then once again. We didn't manage to hit the mamba for quite some time and when we finally did it was not a fatal shot.

I look back now at what I wrote in my diary that night: '*That rock shot outta that catapult and hit the black mamba just behind the head. It didn't kill it, but the mamba dropped out the tree.*'

In real life it all happened so quickly I didn't have time to describe it. I watched in horror and in fright as Reilly, as frightened as myself, forgot to look where he was going and veered into the bank where he promptly managed to get the front of his kayak stuck. The current caught his stern and started to turn it round behind him. The mamba was now headed straight for Rei in sixth gear, but Rei was tangled in the reeds!

The current was pulling me quickly past Reilly; I had only a few seconds to react.

I snatched the *panga* (machete) off its clasps on the side of my boat and threw it in a skim upwards towards the mamba. It was a lucky shot and as it skimmed the water's surface it cut the forward movement of the mamba and made it change direction.

Reilly meanwhile was still trying to

disentangle himself.

The mamba reached the bank and reared up. It angrily struck out in Rei's direction. I gasped; it looked ready to come at us again.

The mamba moved forwards, and for the second time it now showed its soft yellow belly as it rose to an angry head and once again struck out at us.

Now all I had left in defence was my paddle. Reilly was still struggling in the reeds, his face betraying a deep fear. As the mamba launched towards him I hurled my paddle just as he managed to break free. The snake disappeared angrily into the brush and we both drifted quite some distance, pale-faced and clammy.

Luckily the paddle had floated down behind us and the *panga* was close behind it, held to the surface by a float that we had tied to it previously.

When we caught up with Ace and told him the story, he was baffled at our stupidity. Always the sensible one among us, it was Ace who kept us wise down the river.

'Here I am as an equal partner on the expedition, and you buffoons are tempting fate in the very early stages of the river. If you want me to finish up this trip solo, then at least hand over the biltong,' he fumed, 'and your bible, Jimbo' he muttered over his shoulder. 'Seems like I'm going to need that. What the hell are you guys doing?'

Camp was definitely early that night.

Black Mamba Bake

The taste of snake compares only to the taste of chicken and frogs legs; it is smooth and takes the flavour of sweetened wood-smoke with a hint of lemon.

Nutritionally, snake is a high protein food with 93 cal/100g and can be tough and chewy if not cooked well.

This dish is best served as a main meal – for breakfast is best. I would not recommend serving this as a snack, as it may tempt you to rush the delicate process of capture and preparation. There is a line of muscle along either side of the snake's spine; this is the juiciest piece of meat available. The ribs are quite firmly attached to the spine, so scrape your teeth over them to firmly remove the meat.

Serves 3 assuredly mad, but hungry kayakers

1 x black mamba (preferably dead)
1 x fresh lemon (likely carried from the source or raided)
Pinch of herbs and spices
Garlic powder
Salt to taste

1. Very carefully and with utmost caution, catch a black mamba.
2. Using your Leatherman, cut off the mamba's head and hang the body from a tree, tail hanging down.
3. Carefully slit the skin two inches below the now removed head, in a ring around the body.
4. Pull the skin downward and off, as you would do

a sock (or a pair of ladies' stockings), leaving the skin inside out.

5. Now carefully slice down the stomach cavity. De-gut and wash the snake in the river.
6. Rub salt, spices and garlic power into the now empty stomach cavity and place a long green stick within. Tie it with fishing trace.
7. 7. Place the supported mamba above a small flame and rotisserie until baked to perfection. If possible serve with fresh lemon.

The river slowed our minds and our souls and we settled into an easy routine. Here and there, cat-like hippo ears flicked and followed us as we paddled by. There were few crocs but the one or two that we did see were big, perhaps nine feet long. Once we passed through golden tipped rushes, home to a nesting colony of open-billed storks. The fledglings were ever hungry and their desolate calls for 'more food, more food' provided the off-key musical score for the thousands of parents that were taking off and landing in the tall trees all around us.

On an average day we wakened at dawn. One of us would light the fire and put the kettle on to brew. Breakfast could be anything from dry muesli to fish, or sweet rice pudding with sugar and cinnamon. By eight we would have packed away our rag-tag camp into our colourful kayaks and by half-past would be on the river warming in the lemon-yellow rays of the sun.

Elevenses would be spent beneath a shady riverine *Acacia*, hand-lines dangling lazily into the water as we ate biscuits or a small salt-fish fry up. Lunch most days was at one after which we would press on 'till four; we always tried to find camp by four. Five thirty was the latest that we would be in camp fire burning and everything in its place.

Old Mr Connolly from Vic Falls had advised us

that, 'by no account should you leave finding camp until the last moment. When you see somewhere good, stop! Sod's law that if you don't stop you won't find another suitable place for hours and then you run the risk of it getting too late and you'll be wet and cold when you do stop.'

We tried each day to paddle thirty kilometres, but often we averaged only twenty to twenty-five.

Now and again we would come upon natives moving livestock from one side of the river to the other. Their crudely hewn boats carried grunting pigs and silent, arrogant goats that would scrabble about on the floor, giving the natives a hard time in balancing their weights. Cows swam across the river, only their heads and horns above the water.

Every evening our campfire softened the Zambezi river canvas that we lived in and allowed tales and memories of yesteryear to be re-lived.

As I puffed on my biblical roll-up my mind wandered back to a soothing campfire that Dave and I had sat musing around in the Kenyan Aberdares on our search for the elusive Nandi Bear. We had also smoked a small *doobie* to quell the hunger pains; breakfast, we told ourselves, would come with Nandi shaped success.

As we drove slowly up the dirt track, higher into the mountains, deeper into the mist and 'old-man's-beard' that hung like strands of damsel's hair from the cedar trees, the dawn began to lift slightly and we held our breaths. Our eyes pierced through the mist and into the gathering dawn, but we could see little. A bushbuck, its white spots flashing in the early sun rays, bounded across our path and startled us. A lone buffalo bull, chewing the cud as he snoozed, legs curled beneath him, gazed demurely at us, and a small marsh mongoose rose to its haunches to get a good look at us. Breakfast never

did come that morning but the sense of adventure and a brief insight into the dawn life of some of the mountain creatures was reward enough.

These were the memories that collided in my mind as I lay back with Ace and Rei, the Zambezi running quietly beside us.

It was an immeasurable life that we shared then with the water. If a man was seen we shared words, if a bird was spotted we lifted binoculars and if a village was spied we would call in. As opposed to what I had thought previously when imagining our trip, the villages were set far back from the river's edge. Well-trodden dirt paths showed us where water was fetched, typically by the women and children. And rubbish, where there was rubbish there were sure to be humans. We would call at the fish-markets simply to see the species that were on offer. Often we gained valuable information about the downstream river. And everywhere we spoke about rhino. But it was disheartening even at that early stage; the reception we got was one of indifference.

'Rhino?' they asked, 'we have never seen one.'

'Long ago,' one old man told us, 'there were many here but now we do not see them. I do not know where they have gone. I think they have moved to somewhere else in Africa because they were being disturbed.'

One elderly woman described her grandfather and said how he had seen many of these beasts, 'but' she said 'they do not fit here with us now, our grandchildren are progressing and there is no space.'

Conservation is a war of attrition against a constant head of 'progress': development, infrastructure, changing values and the desire of many to live lives as they are lived in the West.

Really does it matter if they never see a rhino again?

Local 'water ferries' plied a trade between the white sand banks. Poled along by a native at the front and one at the back, they sat low on the water, loaded high with chickens, crates of fruit or vegetables, personal possessions and bales of dried fish, as well as four or five people. They were always excitably friendly and although we spoke no similar language we conversed happily in smiles and hand gestures.

At Chinyingi the taut wires of the nine hundred metre pedestrian suspension bridge made the wind whistle as if in salute. Apparently the bridge was funded by a missionary who witnessed the drowning of several people as they tried to cross the swollen river with a sick person.

Many modern missionaries, to my mind, are the death of Africa. Where once missionaries used to live years in remote villages, tending for the sick, befriending the elders and teaching the children, while simultaneously learning the local tongues and traditions and converting precious few, now missionaries seem to be all about fund-raising for two year contracts. Christian volunteer organisations work in regions where the young do not understand customs and traditions let alone languages. In the end the message left behind is folly – 'God will provide.' The villagers quote back to you when you try to ask why they are throwing rubbish in their drinking water source or are wasting water in a drought.

At many villages in this area we stopped to give talks and to show photos. Ace handed out pencils to the school-kids. Reilly spoke about poaching and conservation, population control and preservation of forests and I took photos. At each stopping point our roles would change, but our audience was always inquisitive - only, not about

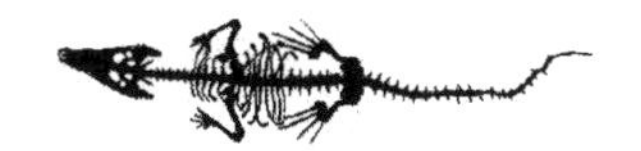

rhino.

We drifted into the small town of Zambezi where the fishermen were bartering their catch for blankets and where the headman marched us down the main 'street' proudly. To our west, somewhere, lay the Liuwa Plains National Park where lions and hyena roamed across flat grass plains, dotted with pools of water lilies and massive herds of wildebeest. But all around us was the sad evidence of poaching.

We had heard tell of the last lioness of Liuwa, a place where once thousands of lions had roamed free across a vast floodplain ecosystem. Lady Liuwa had become famous in her lonely vigil across the Savannah and finally in 2009 two young males were introduced. Sadly though, despite copious mating activity, Lady Liuwa failed to conceive and a viable lion population still looked to be but a distant hope.

Nowadays the successful animals are those that are able to live alongside humanity. "Lady Liuwa has demonstrated over many years that she can live in harmony with human activity, which is vitally important in an open system such as Liuwa, and we want her to teach the youngsters to hunt wildlife whilst avoiding people and cattle."

Of late, several more lions and lionesses have been introduced and Liuwa may once again become the haunt of lions ... but only if they stay away from people! Sadly this now is the task of conservationists, to preserve species while ensuring distance is maintained between wildlife and people, for only then will conservation be compatible with human poverty and population growth.

It was slow paddling as we headed south towards the Barotse. One quiet day, just past Lukulu, having just had a lunch and siesta stop, we came across a man and his son on the river. We paddled up to his dugout, always on the lookout for

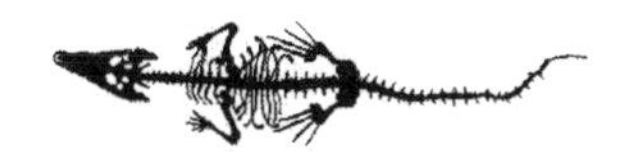

trading opportunities, but the man spoke before us.

'Have you seen a body?' There was no greeting and the man's face was creased with sadness.

The ludicrously of the question at that moment startled us. We asked him 'what body?'

'My father's, he replied.

We answered that we had not and he asked us to help him.

The night before, his father had been killed by a hippo and he and his son were looking for the body so as they could lay him to rest as traditional custom demanded.

The man took us to the scene and talked us through what had happened. There was blood everywhere, together with pieces of T-shirt; we even found a finger. We followed the tracks of the hippo along the riverbank in the downstream direction. The prompt African darkness dropped heavily upon us before we had had any searching success and we spent a cheerless night with this man and his son. They told us of all the people who had been killed in this area by hippos and crocodiles.

That night as I tossed in my hammock, I mused. In the dank night I thought of all sorts of African beliefs, taboos and problems and Tololoshes. The Tokoloshe is a miniature and mischievous water-sprite that is said to haunt certain rivers, the Zambezi included. They are said to hold a certain sexual prowess and often prey on women. Thinking of all their 'badness' I unconsciously curled myself tighter in my hammock. Many of the people in Southern Africa believe that by sleeping with their beds on blocks, this mischievous trouble-maker cannot reach high enough to pull himself up onto the bed. Although, it is said, supposedly a Tokoloshe only becomes harmful and able to talk once he has been captured by witches. The Tokoloshe is terribly

afraid of fire and smoke repels him, so as a result, he lives in lonely places away from civilisation. Round the fire just half an hour ago the man's son had astounded us all:

'I have seen a Tokoloshe' he had whispered to us, peering fearfully into the shadows beyond the fire, 'I have a dog and he can see them,' the son muttered and his father nodded wisely beside him.

'Yes' he said, 'my boy, he took the sleep from the corner of his dog's eye and he was putting it in his own eye.' He explained no further but we knew that by doing this he had harnessed the dog's power of sight. Both men shivered and prayed for us – that we may be kept safe on our journey from the Tokoloshe.

Comfy in my hammock, the mosquitoes whining in the dark beyond my net, I wondered what the chances of seeing a Tokoloshe were. I knew that swallowing a pebble would make him invisible – but I hoped he would like our lonely, wandering ways and would take it upon himself to protect us. I smiled – our protector – a mischievous water sprite. I liked the sound of that. In the stillness I climbed out of my hammock and raised it just a little higher from the ground. *Can never be too careful* I thought. *Better safe than sorry.*

I mused then on witchcraft and totems. All round Imire Safari Ranch in Zimbabwe, we knew people by their totems; those which they were born into. Many Africans believe their lives are controlled by ancestral spirits. These spirits wander aimlessly until they are given permission to come back and protect the children of their own totem. Totems identify the different clans and it is believed that the people display attributes of their totems, be they elephant, monkey or zebra. In this way the clan members are united while issues such as death, marriage and birth will be hugely affected or even

dictated by totems. Totems transcend space and time, and no man may eat the meat of his totem.

The stars twinkled through the trees and clouds as I thought how lucky it was that we were not prohibited from eating fish and how sad it was that we too did not share totems that united us across the world. Finally when I could muse no more I drifted to sleep beneath a scurry of shadow-like clouds.

Morning broke clear and fresh and we again began our search. We did at length find the badly mauled and scarred body and the sons carefully wrapped it for the journey home. I never once saw anything in their blank and black faces, just a silent question; there was no panic or even fear, just the acceptant look of fate that hung on their faces like an accessory for the day. We parted ways with a sound reminder *'that you can come short, and you can come short fast.'*

Some days later I lagged behind the other two and met a fisherman whose tales of past grandeur on the river captivated me. He admitted that all his stories had come from his father.

'Do you want to meet him?' he asked me. I caught up with the boys to bring them back and together we went to meet his father. We hiked through waving heads of bulrushes to a small dry island where three tiny circular huts with neatly thatched roofs stood.

'Our winter houses,' the fisherman told us and bade us sit. He disappeared into one of the huts and we could hear a muted conversation of great animation and soon a gnarled figure appeared at the doorway.

'Welcome, welcome,' a smiling man crunched in a gravelly tone as he crouched to exit through the low doorway. His son followed behind and as they

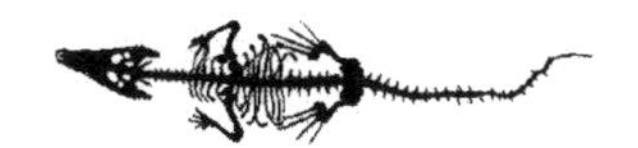

straightened up we could see the son was a full foot taller than his old man whose hair was quite white and surprisingly, because it is seldom seem, the old man had a full beard of white wiry hair that looked windswept, but of which he was obviously proud. As he twirled knobbled fingers around its strands in a somewhat sprightly finger dance, he peered at us through clouded eyes, probably due to countless smoky wood fires within closed mudded huts, and in well spoken English he said,

'So you have come to hear about the time when I was a small boy?'

We nodded and greeted him graciously with all the conversation niceties that are required in Africa – we asked after his wife (who had sadly passed), his parents (his mother passed too and his father taken by a crocodile but the body retrieved) and his children (he proudly pointed across his knobbly knees to his son). We spoke about the weather and the fishing, the cost of living and his heritage. His English was amazingly good and he told us of his days growing up with a good education.

'Better than this government is offering our children these days.'

As we sat and ate ripened mango with golden juice dripping down our chins and heard stories that were populated with cunning leopards, wily hyena and bolshie buffalo, but we were captured more by his talk of burial than of wildlife.

'You white people, you are very strange for us,' he sagely said, his white head tilted to one side, his fingers ever-dancing through his beard. 'I have heard all this talk and now I hear it from you as well, about conservation and protection of things like rhinos and this business of planting trees. But how is it that you are buried?' The question caught us off guard and after a silent pause both Ace and I spoke

at the same time.

'We use a coffin.'

'We are burnt.'

'Exactly' the man said. 'To be burnt you need to burn wood and to make the coffin you also need to chop down the tree for the sole reason to bury it again.'

We nodded, unsure where he was leading us.

'How are your mangoes?' he asked.

'Very good,' we replied, our hands sticky with golden goodness.

'Those mangoes are from my father, come we will go and see him now.'

We followed the old man along his small island, slightly bemused, and stopped before a coppice of three mango trees, heavily laden with fruit, reaching to the cloudless sky at three different heights.

'You see that tall tree? That is my father. When we die, in my family at least, it has become a tradition to plant a tree. But our trees are something special because they are planted above the bodies of those that we love and who have passed. From times long ago the same burial is used. The deceased is wrapped in a curled shape, like a ball, and his favourite blanket is wrapped around him. He is buried in a small hole and a tree is planted. In this way he is like the balls of life-giving dung that are buried by dung-beetles, without which life cannot survive. My father's body and his spirit lives within these mangoes,' and he spread his arms wide and smiled a toothy, effervescent smile that captured us all.

'And this,' his son interjected, 'is my mother.' He pointed to a smaller tree proudly. 'Almost as tasty fruits as my father.'

We laughed and asked about the other tree.

'There are two more trees, but these are

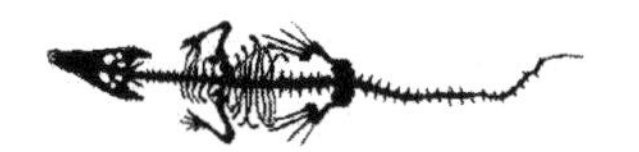

different,' the old man said. 'I do not know how old I am, but I do know that this tree was planted in the year that I was born, and this one,' he pointed at a smaller tree that we had not seen behind, 'I planted this for my firstborn.' We gazed up into the branches of the mango trees and I thought how wonderful this way of thinking was, and how fabulous that the life of a loved one could continue to give such joy and spin such a fabulous tale as that which we had just heard.

"Row Rhino Row" was the name we had given to our expedition in the planning stages, but now it quickly became known as "Roots & Berries" due to the time that we spent rooting about for food and collecting fruits and berries, and each time that we reached into smooth and knotted branches to collect, I thought of the old man and his tale of the deceased and gave quiet thanks to whomever it was who had spurned this fruit or berry tree.

The river's fingers twined around current shaped islands breaking and rejoining again and again; each island seemed to be there only for us. Time gave us languid hours each day to explore and collect, to relax and to fish. We would collect birds' eggs and berries, grasshoppers and edible 'greens'. In the villages we traded fish-hooks for sugar, tea, spices and flaming chillies.

Soon the first signs of the Barotse floodplains began. The trees began to thin out and on the flat floodplains, the nights seemed colder and the sky lower. We passed the Luena Flats region where the bird life became prolific but there were many people and cattle and no mammalian wildlife to be seen. We fished every day, often trawling our lines behind us, and reeled in bream, tiger-fish, one Zambezi parrot fish and what we called purple laviers.

Fishermen all over the world are renowned for their storytelling and we were no different. 'Hey,

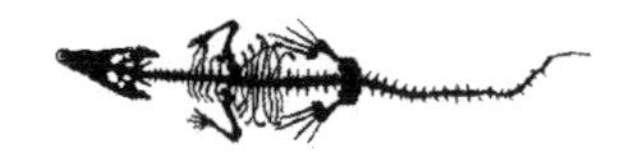

Rei, remember when we were chased by that hippo on that dam in Zim?'

'Yeah man that was crazy – at least we saved the whiskey!'

'What happened,' Ace interjected, 'tell me.'

And together Rei and I relived our night in a tree. It had started as each of our Sundays usually started.

'What should we do today?' I had asked over a steaming breakfast of sweet potato and porridge.

'Fishing' Rei replied. 'Like always!'

And so we had made ourselves a picnic and headed out on the canoe to a sandbank that lay in the middle of the dam. Upon this patch of sand we set-up deck chairs and a cooler-box filled with ice, beers and a bottle of whiskey and settled in for a day of fishing, broken only by our tasty lunch of peppery salami and sun-melted oily cheese in wraps filled with tomatoes and onions, fresh from our own veggie garden.

Soon the beers were finished and late afternoon was pushing back the horizon.

'Time for a whiskey, I think,' Reilly said and rose to prepare. Suddenly a cranky hippo burst out of the water and Reilly started. The hippo circled us slowly and then charged. Its five toed feet propelled it surprisingly fast out of the water and onto the sandbank. In the pandemonium the cooler-box and food hamper were tossed sideways.

'In the canoe, quick! You take left paddle I'll take right' Reilly yelled 'don't forget the whiskey.'

We paddled frantically, Reilly on one paddle and I with the other, bottle of whiskey safe in the boat and the hippo in splashing pursuit.

'Where to?'

'That dead tree – the cormorant tree,' I panted, the exertion and blood alcohol level taking my breath away.

We reached the cormorant tree and ditched all to climb as fast as we could into its bare, white crusted and ammonium smelling branches. The hippo squelched through the mud and belly flopped into a small grassy pool that lay at the base of the tree. Late afternoon turned to dusk and soon the cormorants came home to roost and so began a very difficult night indeed.

Come morning we were shit-encrusted, tired, sore and grumpy and while all the farm laughed we consoled ourselves with the plan to trans-locate the pesky hippo; a tale that perhaps will feature at a later date . . .

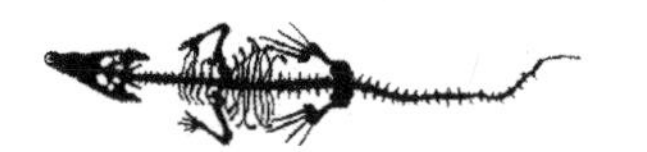

Chapter Seven

Barotse – Sioma Ngonye

Thou shalt take only memories and leave only footprints

It took us twenty days to cross the tree-less expanse of the Barotse. We had been warned it would be easy to get lost here and we heeded the warnings. Each day we navigated with a GPS to plot our route. In high water the river could be over thirty kilometres wide. Without a bearing and GPS, it would be exceptionally easy to drift off course and, instead of travelling downstream; one could find oneself on the opposite river-bank. Often we would feel pleased by our distance, with the GPS recording thirty kilometres, but when we compared it to a map in the evenings we found that we had barely covered half that distance, such was the extent of our criss-crossing.

Sometimes our spirits dropped, especially in the evening as we rushed to eat before the clouds of mosquitoes descended. But later, lying in our hammocks the nights twinkled with stars that seemed so close that I felt I could just lean out and pluck a star from the inky sky. '*I've never seen so many shooting stars*' I wrote in my diary, '*they are like dancing streaks of light that flash across the African heavens. The still water reflects the heavens and we seem so small and insignificant in this vast continent that is Africa.*'

Ace was always our motivator and his boundless sense of humour kicked in at the low

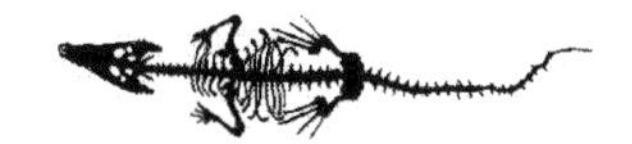

points when Reilly and I were too hot-headed and impatient with fishing or people, distances or plans.

Both Reilly and I knew the places that hot-headedness could lead us to.

I thought back to Imire not so long ago. We had run into trouble of our own making when we had sent a gift to a particularly detestable war veteran. Reilly and I had come home one evening to find that the volunteer house had been robbed; mattresses, food, clothes and fishing gear!

'No way are they going to get away with this!' Rei fumed. 'We're going after them.'

Their tracks led across the farm, behind the dam and under the fence. We picked them up again over the main tar road and followed them right to the door of a house that had been claimed by a motley handful of war vets. We quietly snuck away before we were seen.

Once home we put 'Operation Chameleon' into action. In Africa, chameleons are feared beyond all other living things. Their colour changing abilities make them a creature of distrust and deceit and most adult men are reduced to shivers when they see, let alone touch them. Christmas was drawing close and we knew the war vets would be demanding their 'Christmas boxes'.

Chameleons are easy to find at night because their knobbled bodies do not reflect light but look opaque when caught in the car headlights, so after a few hours of searching we found several poorly fated individuals. We carefully crafted a box into which we packed them, wrapped in silver paper. Come dusk the following evening we deposited this box with a note outside his door. *'Be prepared ... for a new year filled with good tidings.'*

I don't think we had ever considered the repercussions; at least I don't remember discussing

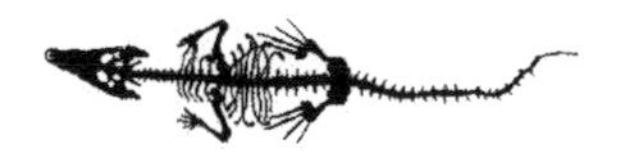

them. It was John Travers, Rei's Dad and the man who ran Imire so well, who saved us and carefully smoothed things over with the irate and quite clearly, terrified, man.

The Barotse yielded no chameleons (that we spotted) and no war veterans ... thank goodness! But we conjured up in our minds the festival happenings that would happen annually upon these river currents.

Back upriver, at the mission in Chitikoloki, the gruff Scotsman had spoken of this festival of the Barotse and the Lozi people, a friendly tribe who called these floodplains home.

'Barotse land considers itself a free state in southern Africa' the Scotsman had said, 'with its own government, defence forces and head of state.'

He reminisced. 'When the high waters are due, the king and sovereign of Barotse Land leaves his island on the floodplains and moves to higher ground.' His description left is wishing we could see the spectacle that his words allowed to unfold before us. 'Dancing men, women and children would parade alongside long black-and-white wooden canoes - whooping and hollering, singing and weeping.'

The Scotsman had paused and closed his eyes to better remember the noise and smells that came with the excitement of the move.

'Huge floats of elephant and water birds are made and stuck onto rounded support structures on the canoes. Men in feathered headdresses and traditional skirts with bright reds and greens, paddle the canoe holding long spears and singing deeply. Behind these be-decked canoes are sub-chiefs and side-chiefs and assistant chiefs and chiefly dignitaries. Crowds line the banks to watch their king move to the higher grounds. They celebrate the end of their fishing season and prey for the fish to

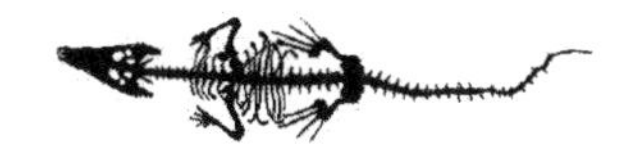

multiply before the season opens the following year.'

We only wished that our passage through the Barotse Kingdom had allowed us to be a part of this fabulous-sounding procession. But we were too late and the King had already departed.

The Barotse kingdom incorporates a number of ethnic peoples. The Lozi people worship Nyambe who is the creator of all things: the forests, the river, the plains and all the animals and fish. He also created man, Kamunu and his wife. As we paddled through the endless channels of the upper Barotse we met a sun-beaten Lozi elder who recounted for us the story of man's waywardness.

'Nyambe, he is our God. Like any real man he has many wives but only one favourite. She is Nasilele.' He stopped and plugged at his handmade pipe: 'how many wives do you have?' His red eyes looked inquisitively at each of us in turn.

'None,' we chorused, knowing that this was the most asked question in Africa. Always the response made us smile.

'Nothing!' the man raised his eyebrows, 'Ah but you must be getting some soon. Here nothing is forever,' his gnarled hands gestured towards the horizon. 'No, nothing is forever;' he repeated and then lapsed again into silence.

Again he plugged his pipe and coughed back into life while we shifted ourselves and lay back in our kayaks, legs stretched over the front, ready for his tale.

'Nyambe, that is God,' he reminded us with a rasp 'had two chief counsellors, Sashisho the messenger and Kang'ombe the lechwe. These counsellors they were doing what counsellors do – they organise words between man and Nyambe. But Kamunu was not a good man - he had many ways that were giving trouble to Nyambe. Sometimes you would see he was killing many animals and even

using poison on fish. Nyambe, he was not liking these new ways and did not know how to make Kamunu change, and so he was escaping to an island with his favourite wife. But that Kamunu, he was not good at leaving things alone. He was liking to pester all the time. Nyambe and his wife, they did not like being pestered and so one day Nyambe took his wife and counsellor, Sashisho, with him across the great river and he went up, far up, to Litooma, which was the heavenly village of Nyambe. To get there, he was having to use a spider's web for climbing.'

The man rolled back on his haunches and dipped one cupped hand into the river; drawing it up to his mouth he slurped noisily as he drank.

'He was using the spider to guide him. But this great God, he could not risk Kamunu from finding him and so he was making the spider blind. This was making Kamunu angry that his God had run away and that he could no longer be spying and so he was trying to build many high platforms so that he could reach heaven. But all of them, they were falling down.'

As he ended, he stared sorrowfully down the river, his fingers twitching.

'Ahhh,' he sighed 'our young boys they are not like they were - they are not respecting, they are only wanting money to buy. Always to buy.'

At this point it seemed like Nyambe (the God) had given up on guiding Kamunu (humankind), and Kamunu was no longer interested in following Nyambe.

We left the old man behind us, waving and grinning.

'You must be getting a wife soon!' he called after us.

As his whistles and shouts faded behind us, another sound buzzed in our ears - a deep hum that

duetted with a whine. As we cocked our heads and looked about us, a shiny new aluminium speed boat with two Africans wearing black leather jackets and dark shades, careened past us, setting us violently bobbing on their wake. Later, when we asked about them, we were told only two words. Diamond smugglers! The old man was right; the modern youth were all about material possessions.

For some nights we camped on coarsely-beautiful sandbanks in the plains but often we were hard pressed to find anywhere remotely dry to sleep. Hammocks were out of the question on these plains as there were no trees and no firewood either. We had brought with us a small gas cylinder on which we cooked our meals sparingly. There were plenty of dry bulrushes that we could use to make low-heat quick-burning fires that helped only to dry our kit. We would spend some nights sleeping on our kayaks or laying our gortex sleeping-bags down on marsh land and waking up sodden, the gortex obviously cheap Chinese copies of the real deal (of which unfortunately the continent of Africa is flooded). The cold nights meant that often we woke up with frost across us, the smell of wood smoke curling around us.

Often the body's response to cold nights was vivid half dreams or nightmares in which the past echoed and other uncomfortable moments were brought back to mind. One recurring dream I would have was based loosely on an incident I had with a lion – only when I dreamed, there were multiple lions and I would awake both sweating and shivering, peering into the blackness for glowing eyes.

I had been offered a job on a game conservancy called Makalali that lies at the Eastern edge of the well known Drakensburgs Mountains in

South Africa. I was to be ready for work on Monday morning. With prior, sensible planning, that meant being ensconced in my allocated *digs* (staff quarters) by late Sunday afternoon at latest.

The party on Saturday night became the pool hangover relief on Sunday morning which in turn became the Sunday afternoon Pimms party. At three I knew I better get on my way – the farm turnoff was a hundred kilometres away and a further twelve kilometres down a dirt, un-trafficked track. I packed my old hessian sack with my belongings (all I had with me at the time) and managed to hitch-hike to the turn off.

By now darkness had settled in fully and the beer, bloody Mary's and Pimms had dulled my senses. I had a decision to make – should I walk the twelve kilometres to the camp along a road that I did not know, that was not only within a game sanctuary that was home to all of the big five or, should I find a place for my hammock and walk in the morning? I was desperate to make an impression and decided that being late for work was not a wise move ... and so I started walking.

All around me the night tumbled with life. Crickets vied with night-jars to back the flashing lights display of the fireflies with a symphony, some form of antelope barked gruffly in the distance, and now and again an African hare rushed into the dim pool of my torchlight. A scops owl *too-wooed* from behind me and was answered by another some distance away. The road was wide and rutted with tall grass verges on either side. As the breeze patiently stroked the heads of the grasses I could see (in my imagination) all sorts of predators. Eventually it was all too much. I needed to sleep. I scouted to my right above the grass verge and found two conveniently placed trees between which I could rig my hammock, and with nothing to eat I

simply took a few glugs of water and settled into my sleeping bag. Within moments I was swayed to sleep by the light breeze.

Suddenly I was woken by a coughing grunt.

Lion!

Silence hung in the air. Did I really hear lion or was it a dream? The grunt came again, a deep *hhhuumpff.* It was unmistakably. Undeniably. A big male lion!

Suddenly the night was rent apart by very loud and very close roar that was answered by another very loud and very close roar.

I dared not turn on my torch. I had not checked if there had been a fence running either side of the road and now I began to worry. If there was no fence then in all likeness the lion would smell me (after all I'd been boozing for a weekend and hadn't washed in a week – Reilly's fault of course!). I scrabbled as quietly as I could into a position that allowed me to feel about on the dirt for pebbles. As I scrabbled, the roars became a definite territorial claim and the noise drowned out my whole world! I could hear nothing but the guttural grunts and terrifying snarls and bellows – cat-calls of lions at battle.

My heart hammered in my chest and my mouth dried of saliva. I began to throw pebbles one after the other towards where a fence should be. *Please please let there be a fence* I prayed. I hoped that a pebble would chink off the chain-link and alert me to the presence of a much needed divider.

One pebble ... the lions tore at each other and panted loudly ... two pebbles and the growls turned acidic ... three pebbles and the air was filled with the sweet musk of fighting cats that turned acrid on the tongue ... four pebbles and no chain link fence ... five pebbles and the fight seemed to be dying down. There was a victor.

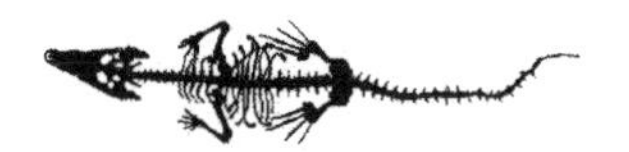

My hands shook and my breath caught, the smell of torn grass wafted on the breeze and the night returned to normal – the cries of the crickets and the night-jars crescendoed with my hammering heart. I fell into a deep sleep once again.

As the strawberry sun laid its ochre over the grasses I saw that there had indeed been a chain link fence between myself and the lions that night ... and as the sun peeped higher she saw me jogging at warthog pace to work.

Back on the river and now about at the mid-way point of the Barotse, we came upon a large and abandoned vehicular ferry that we climbed up onto and explored. Five or six natives were using the vessel to fish from with hand-lines that were tied to thick sticks. The ferry had been donated by an aid agency, they said, for transport in the region. But the region had no roads and the people no vehicles. It seemed to be a white elephant sitting idly on the water, rusting slowly. It represented just another misdirected attempt at salvation for Africa, and meanwhile hundreds of rhinos had needlessly died.

Africa's Western Black Rhino is now officially extinct, having last been sighted in 2006, while the Northern White Rhino teeters on the brink of extinction, saved only by three Czech zoo specimens that have been released into the controlled wilds of a ranch in Kenya. The Southern White Rhino saw a comeback from one hundred individuals at the end of the 19th century to an estimated twenty thousand today. But those numbers are fast disappearing as poachers continue to fulfil market demand. The Southern White Rhino was declared extinct in Mozambique in 2013, while the black rhino has been declared critically endangered worldwide.

South Africa alone saw over a thousand rhinos poached in 2013! That is equal to almost

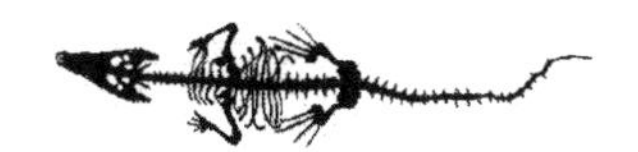

three a day!

Rhino no longer roam the grasslands and marshes of the Barotse and for us the Barotse was a long, wearisome haul. There was a lot of paddling around severely drawn-out meanders. We covered distance slowly but the fishing was remarkable. We caught huge tiger-fish, sluggish vundu and amazing crocs. Six foot long, green-armoured, teeth-gnashing, amphibious lizards from prehistoric times.

Plovers speckled the sandbanks in nest guarding formations that we breached to collect handfuls of eggs, huge flocks of white-faced whistling ducks turned the sky black and tempted us sorely. Most days we found ourselves in pursuit of these ducks but time and time again they evaded our hunts.

It seemed easier to chase these ducks naked as it made for easier camouflage, but then again it was harder to protect the vital bits from thin reed cuts. Just like paper cuts these small scratches would bleed profusely and itch for days afterwards!

Our duck-hunting methodology was sound: Roll in the mud, apply facial war paint and tie grass to your head. The stalk.

We would stalk, deep in the water, only head, shoulders and catapults showing. As we got within shooting range of the ducks we would fire. Our first stone that we fired would send the flocks of ducks into clouds and the second stone, fired randomly into the flock would hopefully hit an unlucky duck.

It was great fun but seldom successful and filled a few hours each week.

But Barotse was dramatically kick-ass. The giant golden-elephant grass spewed forth huge nesting colonies of open-billed storks, sheens of black against the sunlight. Sapphire and indigo kingfishers dived before us on the water, African

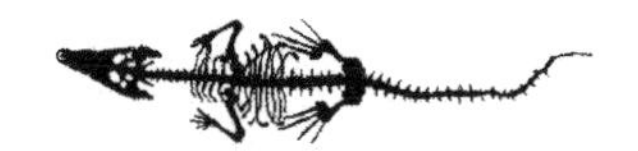

skimmers, their beaks like over sized carrots with yellow tips, seemed to glide across the sandbars, while spurwings, knob-billed ducks and cape turtle doves took turns at calling out warnings. Grey-headed gulls would rasp alongside us, hurrying us along,

'Graaarr' they said as they spun and dipped against the cerulean sky.

For two days we camped beneath three mango trees, the only relief on the otherwise flat plains. Here I managed to badly burn my hand and for two full days I could not paddle at all. Reilly and Ace scouted out some honey with the help of a fluttering honey-guide. I lathered honey across the burn and bandaged it tight. In two days the skin was freshly healed; pink and soft.

Paddling the Barotse was a tough slog but later on down the river we often reminisced fondly about our experiences there: The awesome fishing, sleeping out in the cold nights beneath the stars, the smiling river people that we met, the horrific mosquitoes and the crocodile hunting with our motorbike battery-powered spotlight that charged off our roll-up solar panel. It was quite a scary thing going out at night to hunt crocodiles. Ace classed it as 'Crazy Activity # 101'.

Anyone knows that there is nothing 'careful' about crocodile hunting, but still it thrilled us. Alert, but careful, we waded through the water, waist deep, heading very, very slowly towards a set of orange crocodile eyes. Most of the small crocs sank deep into the water before we could reach them. The young ones seemed more scared of us than we were of them; they would lie still on the river bed or amongst the reed stems until we had passed. The trick was to remember where they had gone down. When we thought we were close to where they could be hiding, we would guesstimate our aim, the

wader, armed with the spear-gun, would stir up the water and as the croc rose upwards it made for a shadowy target. Once it was hit the croc would once again dive deep, but the croc was now attached to the spear which was on a string, which was in turn attached to the spear gun which Reilly, Ace or I held. Now we would have to fight this croc, bring it in, hold it down and quickly kill it.

It was the bigger crocs that worried us. They were not frightened and their orange eyes watched seductively until the spear was fired and the frantic splashing began. Then they would disappear, melt into the liquid blackness. Of the three of us, one would wash the torch beam over the surrounds while simultaneously thrashing the water with a paddle until the kill was made and we were 'safe' in our kayaks.

Were these night forays wise or safe? Inasmuch as we could make them they were - but in the end luck held with us - would I do it again? Absolutely, even though I know that eventually nature will always win. It is the feeling of being part of nature: even if the hunter can quickly become the hunted, the exhilaration is life giving.

Beginners make mistakes and on our first dinner of crocodile we made a terrible mistake. We had heard that if the bile from the gall bladder taints the meat it will poison it. But this pearl of wisdom was in the recesses of our memory. We planned croc tail for dinner. The body, we would gut and salt in the morning when it was light. In removing the tail Reilly cut too high. I took the upper tail end and Rei the lower, Ace turned down the offer of fresh crocodile meat, he already had a 'jippy' tummy from something we had eaten a few days earlier. Reilly and I ate well.

Within a couple of hours I had to make frequent dashes to relieve myself. This in itself was

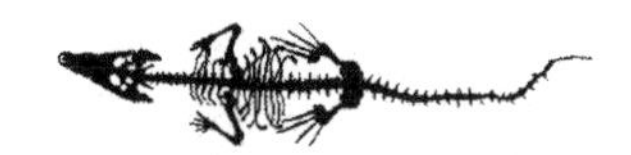

not easy on the cold night. We had no toilet paper and so had to go in the water which was a scrotum-shrinking cold and meant precious minutes by the fire trying to regain warmth. Not to mention attracting the fish and feeding the crocodiles!

Coal Cooked Crocodile Tail

Of the crocodile meat there is no better alternative, a succulent white meat that is delicious and unique. A meat that is low in fat and high in protein.

Self caught crocodile is a 'man's meal,' a dish that proves his strength and worth in the race of survival of the fittest.

This dish is best served as a main meal for a late dinner as the process of hunting for crocodile is long and dangerous. Once you are warm and dry and safely eating this delicacy round the campfire you can celebrate your 'aliveness.'

Serves 3 wild and daring but *still* hungry kayakers

1 x 6ft crocodile, tail of
1 x fresh lemon
Pinch of herbs and spices
Pinch of salt
Pinch of chilli

1. Carefully hunt and catch a six foot crocodile.
2. Using your Leatherman remove the tail below the gall bladder (two inches below the ass hole).
3. Do not skin the tail but bury it in a mound of hot coals and leave for twenty minutes.
4. Remove the charred tail and cut it open latitudinally.

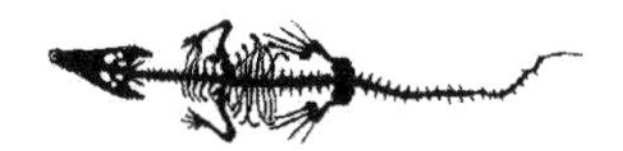

5. Open the tail, almost like a book, to a delicately white BBQ flesh.
6. Add a squeeze of fresh lemon, if you have it, salt, herbs, spices and lots of chilli.

I woke in the morning to deliver the news.

'Boys, I'm in a bad way. I've been in the bush all night.'

That day I tried to paddle, but with the hot sun and my constant getting in and out of my kayak I soon lost strength and we had to have two days off while I recovered from the most horrific stomach cramps and diarrhoea I have ever had. I was severely dehydrated and it took a full week for me to regain my expedition strength.

Reilly didn't waste this downtime that we had. He had decided that he needed one of the teeth of this crocodile to give to Candice, and so he set up the *sufuria* (saucepan) on the sparse firewood, and tried to boil the skull down so as he could extract some teeth, but as with most of Rei's escapades it went wrong and the skull fell un-noticed into the flames and was reduced to ashes; crocodile ashes.

Hilariously though, as we drifted down-river on that day after the crocodile bile poisoning incident, I was trawling my line and managed to hook a substantial tiger-fish. I needed desperately to head to the bushes and Reilly and Ace were both there, laughing and goading as only boys can. I jumped out my kayak and managed not only to achieve the toilet stop but to reel in the tiger-fish too!

Falling so ill while travelling is never fun and I was reminded of a malarial incident back home in Kenya. Malaria is the dreaded malaise of all travellers to Africa. It is common but it can be lethal.

I had covered over a thousand kilometres

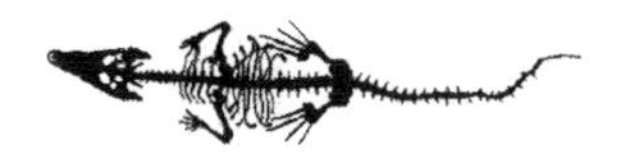

hitchhiking back to Kenya from Zimbabwe. At the Kenya-Tanzania border I felt the beginnings of the lurgy but being so close to home I believed I could see it through. Big mistake! By the time I had reached Naivasha I had not a shred of energy left in me and in my delirious state I called for the driver to let me out. Then there is a blank in my memory. I must have been left on the side of the road with my backpack. I did not even make the much needed shade before I collapsed on the verge in cold sweats beneath a scorching sun while rubbish blew in sweltering whirly-gigs around me. Sometime later I was brought round by a face that swam in and out of focus.

'Jamie Manuel?' it queried.

I could not summon a single word to my tongue.

'Well you look like him and I'm guessing you're pretty sick. Malaria it looks like.'

I was loaded into the car and the next I remembered was waking in a closed, dark room where the nightmares began; illusions of charging pieces of cotton being ridden like steeds through the air by men-of-war while water was forced upon me. I lost a whole week of my life then but the body seems not to remember such pain and anguish and as the crocodile bile fought to poison me I could think only that I was on the verge of collapse.

Crocodile bile and malarial tales aside, our Zambezi paddling allowed us to meet many fine and interesting characters. At the end of one long day we met Simeon, a wrinkled old crocodile hunter whose hair was quite white against his black skin. He regaled us with tales of old. He remembered the herds of elephant that would come down into the swamps in their thousands, the huge herds of blue wildebeest and the lion that used to haunt the fishermen at night.

'Back in the day,' Simeon said, 'we would shoot about ten or twelve elephant a year. We would sell or barter the ivory and eat the meat. In this way we improved our village lives and planned for the future of our children and grandchildren.' Simeon gestured for a cigarette that he shakily lit before he continued.

'I do not remember killing a rhino because for us there was no use. And then things changed and gangs with guns came and took our wildlife.' He sighed, 'and our children too. They learnt how to use weapons and how to make quick money from the elephants. And we learnt too that the price of rhino horn far out paces that of elephant and soon we had no rhino left.'

We collectively sighed as he continued, 'but it is not our fault. It is the fault of you people from across the oceans who want to buy such things. You come here and tell us we should not kill but for us, that is saying we cannot earn money. We kill because you desire.' He puffed at his cigarette almost violently and then he spat upon the ground and signalled that he must leave.

As more and more trees appeared on the horizon, we knew our Barotse days were ending. The town that marked the end was Senanga and by chance the following day was Ace's birthday. We found a little campsite where we pulled in, unpacked and ordered food at the fresh looking, wooden camp-site bar. Two plates each and then went out on the town. We played pool, visited the local *shebeens* (bars) and swapped stories with the natives. Ace had a very unique birthday.

The next day we went shopping for supplies and as his birthday present Ace bought a colourful blanket that he made into a giant double hammock. The luxury proved to be too heavy and the hammock had to be ditched, floated forlornly off

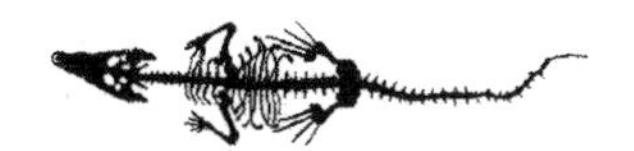

down the river towards a fishing village. Its wet weight was just too much to carry.

Ace's best present though, was that the monotonous slog through Barotse was over. The days of paddle stroke after paddle stroke, those days of endless flat lands and bull-rushes where our tired eyes rushed to focus in on anything over two metres were done. At the campsite we met a friend of Ace's who had brought a further supply of biltong (very important) and a new tracking unit (less important). Plans were that friends and family would be able to follow our movement live and online by way of our tracking unit, but as with many plans in Africa, this one failed miserably.

As we left Senanga the river banks became accessibly lush and fertile. *The 'river-bush is full of bird life and butterflies dance to the evening sun. The people here are full of life and smiles.'*

Further down-stream there were islands that rose as wooded knolls from the water, where huge trees spilled green over the edges, while gregarious palms rasped in the breeze. Here we spent extraordinary evenings round the campfire. Reilly's favourite occupation was setting and baiting the night lines. Often we fished only for the interest of which fish species there were and released our catch as soon as we had identified it. '*There were jackal berry trees, lush rain forest like undergrowth, Riffa palms and emerald ferns*' around every river-bend. Slowly, slowly things were getting quite wild. Basks of crocodiles lounged with open mouths on the banks, big groups of sun-wary hippos hung in pods, but they were nothing compared to the numbers we would soon see.

When we arrived within the growling sound of Sioma Ngonye falls we decided to pull in and rest

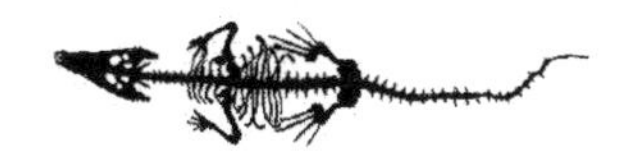

before the morrow. We had no idea how big the falls were or what they looked like. The river was a maze of channels that we knew nothing about and that were not marked on any map that we had. That night we sat up ruminating how we would know which channel to take. Not many fool-proof, stone-proof plans arose and when we turned in that night my mind was buzzing with what lay ahead. I had pulled the short straw. It was me who would lead the charge down these falls tomorrow. My imagination was alive with all sorts of scenarios.

The next morning we put on our life vests and spray skirts for the first time in ages. We pulled out into the current, and I took the lead followed by Ace and then Rei. The river got faster and faster and I was constantly choosing the channels based on nothing more than what I could see. We hummed down small rapids and then down more.

Suddenly we heard faint calls behind us.

'NO, NO, NO!'

We cascaded into a small eddy one after the other and held ourselves on the roots of a massive fig tree. We could see no-one.

Soon a fisherman who spoke good English materialised out of the bush in his dugout canoe, nosing towards us on a ribbon of current, his face shrouded by hanging lianas.

'Ah you must not be going that way or you are going to die' he calmly said.

Kevin was a first-rate guy who had learnt English in school and who seemed to register no surprise at all in hearing what we were doing, or in seeing us heading down blindly through his fishing grounds towards a raging torrent that we could quite clearly hear.

'Follow me' he called, and leaving his dugout tied to the fig tree he helped us push and pull our kayaks back up some of the channels we had just

come down. Eventually after three or four hours we got to the top of the rapids and he showed us a side route down a relatively calm stretch which allowed us to pull in to the bank in an eddy at the top of the falls. By this stage quite a few fishermen had gathered around and for five fish hooks each we negotiated portage around the falls.

Sioma Ngonye falls were an awesome jumble of sharp basalt, black as night. The falls fell over a staggered twenty metre drop over the deeply potholed basaltic rocks. As we stood, taking in the sheer volume of water that passed over the falls, we could feel an underground flow pulsing at our feet. Here the Zambezi River leaves the Kalahari sand floodplain and the falls mark the beginning of the deep dykes that eventually become the tremendous gorges of Victoria Falls. *Mosi-au-tunya*; the smoke that thunders.

All around were black boulder banks and islands that were thrown over with clean white sand. Skirting the falls had not put us beyond all risk yet. The rapids below the falls were the biggest we had yet faced and people had spoken of them with such vivid descriptions that had already built up a terror of them in our minds.

We scouted the rapids, looking to see which line was the safest to take. What surprised and confused us, was that the rapids did not look too hairy at all. In fact they looked happily runnable. And so we launched our kayaks, expectant of the worst but hoping for the best. Both Reilly and I pushed through each wave and around each hole with only a few wobbles while Ace managed, quite amazingly, to whizz down them backwards without capsizing.

'Quite smooth really!' Rei called.

Ace nodded and pointed, 'especially knowing that those dudes were watching us from less than

ten metres away!'

Rei and I glanced, there lay three or four large crocs on the river bank, months open to the sun and another three that we could see in the water!

And so a section that we had been worried about for many nights had been completed with no mishaps or crocodile encounters what-so-ever!

(We later found out that the water levels had been so perfect as to allow three novices to run a set of rapids that were usually pretty difficult).

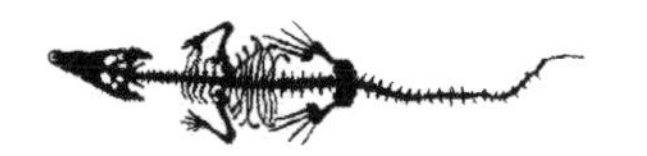

Chapter Eight

Sioma Ngonye - Goma

Thou shalt learn about all river gods – be they women or myth

Maziba Bay. A contrast of colours, a whirlpool of riverine sounds and a time-stopped inlet of beauty. The sun set with a fire that singed the clouds a flamingo-pink and the pure white sand squeaked between our toes as we dug them deep to absorb the last vestiges of warmth. Summer was on its way - just thinking of the warm nights thrilled us. As we lazed about in our chosen river camp the bronzed smoke from our *mascottis* curled into the air. Our minds wandered freely across the view before us, leaving no footprints and taking only memories.

Humongous rocks, rocks that stood on a shelf above the white sands, warmed our backs as we lay back and listened to the sound of the water. Night brought the sounds of splashes as fish leapt away from crocodiles that jostled for fishing grounds. The sands below would soon be prime nesting territory for the female crocodiles as they came into season and began to lay their eggs. In a month or two, these sand banks would be 'out-of-bounds'. The campsite was so exquisite and calming on the soul that we stayed a couple of nights. The fishing too was incredible; tiger-fish that never gave up the will to live flashed silver scales at us as they fought the line, and gigantic bream seemed open-mouthed with despair as we reeled them in.

Lying spread-eagled on a massive boulder, one of the many that dotted the white sands of

Maziba, my mind wandered back to a night I had spent on the Indian Ocean, my toes also dug deep into white sands, beach sands strewn with shell fragments washed in from the ribbon reef that runs down the extent of Kenya's coast.

Lying just North of Manda Island, Lamu, close to the Southern border of Somalia lays the Boni Forest. This forest is little known to any but intrepid travellers and is a quiet and tranquil place of endless beauty.

A friend and I had come in for a couple of nights to explore the forest. Each dusk we lay in the warm sands recounting the day's findings over a cold Tusker. (A Kenyan beer named after the founder of the brewery, a man who had been killed by a big male elephant known as a Tusker). We let silence hang in the air as the butterflies spluttered and hung over the breaking wavelets that ran up and down the beach. I watched these butterflies with a keen interest as in my spare time I am a lepidopterist and was looking for a fine moth specimen that I wanted for my collection.

As we lay we listened out for the quiet sounds of heavy footprints that would tell us the elephant were approaching. We had been told these grey giants wandered along the white sand beaches and played in the ocean's salty water, a sight that is now seldom seen in the vanishing wilds of Africa.

The low tide waves played chase up and down the beach and turacos called as they settled in to roost for the night. The golden orb of the sun was sinking ever lower when five honey badgers appeared unexpectedly on the sands below us. Honey badgers are renowned for their daring and fearsome attacks. Now they were calling to each other with soft squeaks as they scampered about on sturdy legs with wide feet that sank deep in the still wet beach sand. Their broad backs ran up to

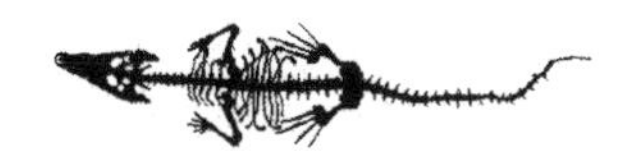

massive heads with short powerful jaws and small dark eyes. With their long claws they were breaking open mussels to lick out the tender flesh within and were galloping in slow, almost clumsy lollops after skittering crabs that they grabbed quickly and shook, like a terrier shakes a snake.

Amongst the five was a youngster who had not quite harnessed the skill of catching and shaking crabs. He ran with a comical pigeon-toed trot and a rolling gait and would constantly halt his search for food to dunk himself in a saltwater pool and then roll about happily in the sand. At one moment a crab latched onto his nose which enlisted a whistling-hiss and a rush to his mother for help. We watched with smiles on our faces and spats of laughter as the tide rushed back and pushed the honey badgers home into the forest for the night.

Rei shook me awake.

'Wake up Jimbo, we need to get cooking. Ace has hooked us a good sized tiger.'

I roused myself from my now cooling boulder and joined Ace and Rei at the fire where we roasted our tiger and then slipped into warm sleeping bags as the night fell about us.

When we awoke the sky was painted in such a perfect blue that it appeared like a deep lake from the tales of Arthur and Guinevere. A perfect day for fishing! We lazily fished from our kayaks, legs sprawled over the bows, the sunshine chasing the last shivers out. It was still early and mist rose off the water; Ace had already caught a fish or two and Reilly and I were in line and champing for our first morning catch.

Suddenly BOOF!

We all jumped.

To my left *'a volcano of water blew up out of the depths and created a massive hump, only metres*

away. Had I been any closer I'd have been flung in some crazy direction and without deck and paddle I'd most certainly have swum.'

The noise of the blow-out itself was enough to cause alarm in the troops, and we watched open-mouthed as this explosion shook the gorge.

Water blow-outs are caused when large volumes of air are forced down by some deep running current that sucks the air into big balls of bubbles. The churning air-balls can only remain trapped for a finite time before they force their way to the surface and erupt in spectacular moments and terrifying fury.

I had been frightened out of my wits once before by a mass of water and this Maziba blowout brought back to me the memory. Below the vast Mutemwa Rock in Zimbabwe, which we often climbed for sundowners, was a large dam and we had taken out a small rowing boat. The plan for the morning was spear-fishing. We would take it in turns, an hour each, beneath the inky, cold waters to see how many tilapias we could shoot and bring in for trading.

As our little boat bobbed on anchor I noticed that the sky was darkening.

'Think there's a storm on the way Rei.'

'One more – this is my last dive, it's getting pretty cold anyway,' and he disappeared, dipping below the waters once more.

I set myself up for a small nap and dozed quietly, wrapped in both of our jumpers when suddenly I heard a strange *whooshing* sound. I opened my eyes and they must have rounded in surprise as just a hundred metres away, a water spout, about a foot wide, was curling into the air, just the way a dust devil does. Before I knew it this whirling tower of water was upon me and much,

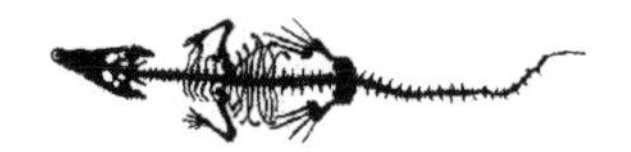

much larger. The little rowing boat was lifted into the air a good two metres as I was thrown violently from it. The row-lock splintered and one paddle flew past me as I hit the water, the boat coming down a good three or four metres behind me with a crack that brought Reilly up from the gloom.

'What happened?'

'Look!' I cried as I pointed behind him to the quickly diminishing water spout.

The whole event must have lasted no more than thirty seconds and yet the sheer force of the water was incredible. One paddle was taken over fifty metres away and across the waters floated all of our day's catch.

The blow-out on the Zambezi did nothing except play with nerves and chase us to the mouth of the bay where an old tourist lodge looked forlornly down at us, abandoned and lonely, needing only human company to come alive.

Time and the water's flow pressed us on down towards Sioma Gorge. The night before we entered the gorge we were kindly invited to camp in the garden of an exquisite house where we were able to use the outdoor barbeque and to sit on wooden chairs with comfortable cushions looking out over the water, a luxury that allowed us to pretend at being kings.

In the morning when we woke, a mist hung over the river and took an hour or more to clear, while through the mists we could hear a pod of hippos grunting blissfully in the dawn.

As I sat on the bank waiting for the mists to clear I recalled a breaking dawn on the Meru River in Kenya with Lovat. As the sun fought with the morning mists (a battle they were sure to lose up here in this arid region), lions '*hrrumffed*' from far

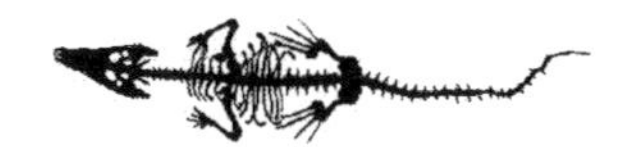

away and the rasping call of the ibis peeled across the waters. This was real Africa. The Africa of vast sand rivers, waving doum palms and giant herds of elephant. Suddenly without warning a massive elephant appeared out of the mists headed straight for us, its undercarriage dripping wet with brown, muddy water. We just had time to hunker low and the bull passed within ten feet of us, seemingly never smelling us and certainly never having seen us.

We let our breaths go as hippo called their grunting condolences from the river.

The Zambezi River mists cleared and Sioma Gorge came into view. Sioma Gorge was home not only to honking hippo but fish-eagles too. There seemed to be a mating pair every few kilometres. On the Zambian side of the river they called to us from the roofs of abandoned lodges - *'Come - here, here, here'* - and we obliged by stopping to poke about.

There is always an incredible feeling in walking about abandoned buildings. In the lounge of one dilapidated lodge there was fresh spoor from a big male leopard and in another the baboons had made themselves at home and had wreaked havoc on the fine woodwork and wall designs.

It was sad to see these derelict, sun-baked buildings crumbling against the bases of bare, thorn scrubbed gorge walls. It was easy to see that in the very recent past these places had been sanctuaries on the side of one of Africa's finest rivers.

We never did find out why tourism on this, the Zambian side of the river, was so much in decline, why the tourists had fled when the tour-operators had closed down.

Further down the gorge we saw duiker and reedbuck and heard the saw-like call of leopards

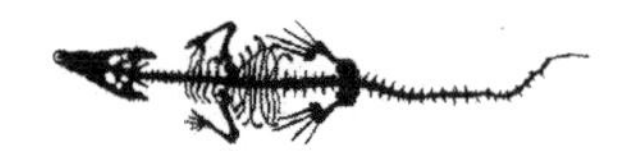

night after night. Polka-dotted flocks of guinea fowl tempted us from the river's edge. Each evening we tried to shoot a guinea from their roosting trees but with no success. They were too quick for us and our catapults were too rudimentary.

The rocky gorge walls were of drab browns this late in winter, but we could see the huge grey trunks of tropical trees struggling upwards from the parched undergrowth, waiting, waiting for the rain, so that they could flourish in green.

One fish-eagle, its pied colours bold and haughty, swooped down on a fish that I had caught. He wrestled it up to his rock perch in his claws while water droplets hung silver in the sunshine.

Ace and Reilly had sent messages home to let family and friends know the dates that we expected to be in Vic Falls, and now those dates pressed upon us. If we were to keep time we would have to night-paddle which was something we were not keen to do.

But as they say, desperate times call for desperate measures, and we were young and bullet proof and capable - or so we thought!

We longed for the reprieve that Vic Falls would offer us and our aching backs, our bruised legs and our beer-thirsty taste buds. We had been warned umpteen times about night-paddling. It was dangerous as it gave crocs a wide advantage, we couldn't see where we were going and we didn't know what was ahead of us. We had to rely not on sight, but on hearing.

Looking back, I can see it was in such moments that our inexperience shone through.

We put out onto the water, the night pushing up against our comfort zone. Snapping our spray-decks onto our cockpits we slid over the river pebbles and nosed into the flow. Instantly we were

swept away – our kayaks and us – into the unknown, only our glow sticks allowing us to keep track of each other.

It was eerily calm. The weak moonlight glanced off the water and our paddle strokes sounded loud and comforting in the vast inky blackness that surrounded us. Now and again our torches picked out the orange eyes of the big aquatic lizards, the ubiquitous crocs of the mighty Zambezi.

We had pushed some fifteen or twenty kilometres before we heard the sound of rapids. The sound was not intimidating - a soft roar that sounded not too scary.

Approaching unknown rapids at night and thinking we could run them - we were bold and stupidly oblivious to danger! We were summoned by an urgency that existed only in our own minds: rest and recuperation, a week of girlfriends and varied foods.

Approaching the section of the rapids that sounded weaker, we pulled in as a tight bunch of three, peering forward into the darkness, trying to make out the churning water ahead. Our hearts were pounding and our stomachs were knotted with apprehension. A slight taste of fear tingled in our mouths while our eyes were anxiously assuming and questioning. *Actually this was pretty scary. Who said anything about night paddling being fine? What lay ahead? Would we make it through?*

The current pulled us downward towards a huge black shape that I glimpsed in my peripheral vision ... and it was drawing closer quickly.

Thirty metres –

Twenty metres -

Ace had seen it too. 'What's that?'

'Shit, it's a monster of a hippo. Shit. Shit. Shit!' Reilly began to panic as we were pulled closer

still. Suddenly I saw another shape just metres away from the first.

'DANGER!' I called 'DANGER! DANGER! DANGER!' I gasped as I realised how close we were to the end of our lives.

'Guys back paddle. See the hippo? A baby, it's on the rock, eleven o'clock.'

Back paddling didn't achieve much. We were held firm in the swirling current. Only a short time before, this baby hippo must have been caught in the very same current, and washed down the very same channel as us managing to get its footing on some rocks mid-channel. Its mother was making her way downstream for the rescue of her calf and we were hot on their tracks and couldn't stop!

It was not a good situation.

Now both hippos were standing on a rock in the rapids. They seemed too nervous to move into the current and the mother was not going to leave her baby.

Fifteen metres -

Ten metres -

Five metres and we managed to make a hard right turn down a small side channel.

As we pulled from one current to another I saw Ace wobble and begin to go over, but I should have been concentrating because I too suddenly felt my kayak lurch unsteadily!

Ace braced. I braced. We were safe.

Our hearts racing, we pulled into an eddy and caught up with our heartbeats.

There was nowhere to pull out.

'Carry on.' Rei said grimly. 'We have no choice.'

Ace led us and Rei held up the rear as we followed the river. A hundred metres or so on and more and more channels started appearing. Debris and undergrowth grew into the channels and it

became difficult in the relatively fast current to navigate our way.

Suddenly we heard the sound of a stronger rapid and before I knew what had happened Ace went over.

I remember hearing his wild shout:

'Help!'

We frantically paddled closer, the night too dark for us to see clearly what had happened.

'Shit,' Reilly said again 'Ace, are you OK?' Only a tiny bit of his kayak was showing, wedged in among logs, vines and rotting leaves.

No answer. Our torches caught the eyes of spiders as we swung the beams over his boat, still not quite sure what had happened.

Reilly started shouting.

'Fuck, Jimbo, we gotta get him out, he's been under already for at least twenty seconds.'

Both of us were already tying our kayaks up and leaping out into the shallower water on the channel edge. There was no movement from Ace's kayak at all.

We fell into the water in disarray and had to clamber to hang on to protruding roots in the current.

'Jimbo, pull yourself onto the bank,' I did as Reilly said.

I could see now what had happened; Ace had been T-boned against a fallen log, upside down for how long I didn't know!

We manoeuvred ourselves, Reilly in the water and me on the bank, and by heaving and pulling we managed to dislodge Ace's kayak from beneath the log it had horizontally trapped itself under, enough so that he could frantically gasp breaths of fresh air. It was a full five minutes before we managed to get him out. His face was ashen and his body bruised.

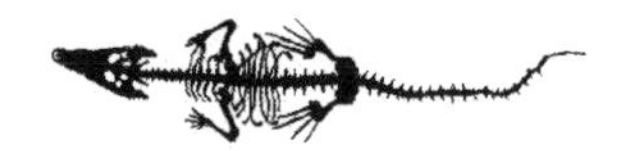

He slumped, shaken and heaving laboured breaths in and out while trying to restore his own confidence.

We were all shaken but where we stood, amongst the debris of the accident, there was nowhere to camp. We would have to carry on in the hope we could reach a decent campsite.

We were in exactly the situation we had been warned to avoid!

And so we forced ourselves to collect all gear that we could see and to climb back into our kayaks. Ace braver than both of us. Luckily we found dry land with space enough to camp within minutes, and now had to start building a fire to dry our sleeping bags and hammocks.

We quickly made ourselves mugs of sugary black tea. Ace was still in shock. We all were. We stoked up the fire to medium-bonfire size and hung our things to dry around it. We followed tea with warm cup-a-soups and assessed the breakages.

We had managed to destroy quite a lot of our equipment. The roll-up solar panel was ripped to bits, two of our four torches had gone down the river and all our equipment had gotten wet.

'Video camera?'

'Gone.'

'Ace, your spray-deck is ripped too.'

'I know' Ace looked morosely at his deck, 'but I'm alive' he muttered into the steam from his soup.

'Yeah, pretty lucky.'

It was five in the morning by the time we got to sleep, and when we awoke we could see that Ace's kayak had taken quite a beating. Quite a lot of food had gotten wet and was lost. But we were alive and well and had learnt a valuable lesson from our gamble; although we still did a few night kayaking stints, we always reccied ahead if we heard the tell-tale roar of a rapid!

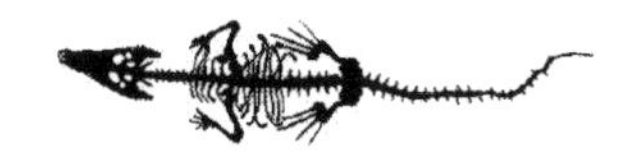

This was the story we had fresh on our minds to tell that very afternoon when we came across the managing director of one of Zambia's monetary banks. He and his wife were building a retirement house on the river. They were not impressed by our story one bit and invited us for a recuperation dinner – a much needed break from our third grade rice and fish.

As we talked that night I remembered an incident that had occurred when I was at the tender age of fourteen, perhaps the first un-scouted rapid that I had run.

Outdoor-loving fourteen year-olds are prone to chaotic plans that they believe they can pull off. I had spent all day making a narrow raft from wood and bamboo, forest vines and sticky sap. The plan was to run a stretch of stream from Monkey Bridge to Lake Elmentaita.

'Not a chance!' my mother said, 'how will we know if something goes wrong?'

But in the end we persuaded her and the next morning we lowered the raft from the bridge and climbed on: I and an equally skinny, adventure-loving kid called Warren.

'*Yaaar* here we go,' roared Warren, in pirate mode, and we let go of the planks of the bridge never thinking what may be facing us.

Only a hundred metres downstream a thorn tree had come down across the flow, its sadistic branches reaching down into the current especially to catch young adventurers. As the tree loomed closer we looked at each other wildly, our only option seemed to be to ditch the raft and swim under the tree. Clinging to the raft, we would be too big to fit through the small gap that was left to us and the current was too fast to allow us to stop in time. Warren leapt to the left and managed to catch

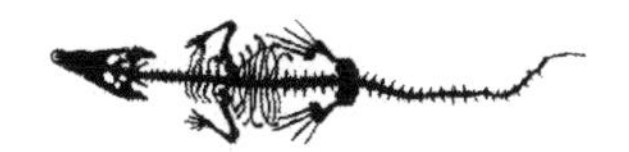

hold of a protruding root but I was too slow and was swept from my feet and into the river, where, open-eyed, I tried to swim deep. Suddenly a searing pain went through my eye and everything went black. I came up on the downstream side of the tree clutching my eye and yelling for Warren.

He appeared from the undergrowth having managed to avoid being swept under the malignant tree trunk.

'What's wrong?'

'My eye, my eye,' was all I could say.

Warren paled and decided that we must go back to camp. 'You'll have to get your Mum to look at that,' he said, and so we traipsed back to camp, tails between our legs and my hand firmly over my eye.

Mum was not happy. Deeply embedded in my eye was an acacia thorn that had to be extracted carefully with tweezers and washed with stinging salt water.

'I hope this isn't the way you mean to go on?' Mum said as she patched my eye with clean bandages. And here we were, nine years later, having just run an almost lethal stretch of river in the dark.

As we wound our way to the end of Sioma Gorge on the Zam, we came upon knolls in the river that were a tangled mass of huge trees. Their great branches rose from close to the earth and extended outward reaching for the gently flowing river. Their leaves were a vivid living green that contrasted greatly with the grey winter thorn scrub that grew just back from the banks.

Once or twice a few incestuous monkeys scampered up an island bank, a baby or two clinging to its mother's stomach. Birds glided and soared, dipped and twirled above and around us. Little

kingfishers called merrily from leafy overhangs and water monitors swam quickly away. For a couple of hours we wandered among things like these, bird-book and binoculars in hand, while we shouted out the names of species, one after the other.

These leafy islands quickly earned the nick-name 'Paradise Islands.' Few of them looked like they had ever been trodden on, while others seemed to have private residences. The current drifted us past an island which looked lived on and loved.

'Hey boys, we need to do some charging, my camera battery is dead and the old sat phone needs a bit of juice. You think we should go ask at this place?'

Ace, of course, was thinking ahead. We both agreed with him and paddled to the base of the stone steps that had been set into the side of the island, and that ended in what looked like a new jetty. Tying our kayaks up, we pushed Reilly to the front and headed up the stairs. Three skinny, sun-browned lads all in need of a good shave. The pebbled pathway led deep into shaded vegetation and to an office that was tucked away under some large, long-legged Riffa palms.

'*Hodi*,' we shouted, the polite Swahili word that lets someone know you have come by on a friendly visit.

'*Karibu'* answered a soft voice from beyond the shaded interior. You're welcome.

The angel who came to the door blew us off our feet. We felt love at first sight for the slim, angular faced Dominique Smith. All three of us were stuck for words. In my diary I waxed lyrical about her wild beauty. I especially was sun-smitten, and Dominique at that moment was my sun. It turned out that this island belonged to a man called Bruce, who was immediately dubbed Bruce Almighty. Dominique was the family au pair.

The family was thrilled to hear of our expedition and invited us to stay. That evening we ate outside amongst the tangled forest listening to the off-key orchestra of cicadas. We shared adventure stories and anecdotes, and in this way we whiled minutes into the dark hours and it was almost dawn when we happily slumped into our hammocks. Dawn broke forty minutes later with the fullest and most deafening morning bird chorus that I have ever heard, before or since.

Electronics charged, we waved goodbye to our Paradise Angel and the family she worked for, and pushed out into the Vic Falls bound current. The boys teased me mercilessly about Dominique and as I know that they will be reading this, I wish to include another 'girl' story.

Not so very long ago I had a girlfriend who lived and worked in Nairobi (Kenya) at the tented camp that lies within the National Park. I would drive to Nairobi from Kora often reaching the city on dark, well after the national park gates had closed. I'd ring Kim.

'Hey Hon, I'm in town. Can I visit you? Where are you now?'

'Ahhh Jamie I'm at work and they won't let you in now, I'm afraid. Plus I've a load of guests tonight so I'm pretty busy. Let's meet up tomorrow morning.'

I hung up and debated. I'd driven all this way especially to see her and I'd be dammed if I'd be stopped. I hired a *boda-boda* (motorbike taxi) to take me to a quiet part of the game park fence, where I paid him off and waited for him to disappear. Then up and over the fence I'd shimmy.

Now the camp that she worked at was some five or six kilometres inside the fence, through a section of forest, past a rock *Kopje* and down over the Savannah to a small forested glade. The first

time I did this walk, in the dark with a torch that I could only shine intermittently so as to avoid being spotted by patrolling rangers, I was slightly nervous. Soon though I knew the route and was more at ease.

I would surprise her by being in her tent when she got back from saying goodnight to her guests! I could expect to pass buffalo and giraffe but when I happened across a leopard, both he and I were startled.

'*Whoosssh*' I breathed out slowly and carefully. 'Now Mr. Leopard,' I coached in a tone just above a whisper, 'all I want to do is see my girlfriend, so you go this way,' I gestured right, 'and I'll go this way,' I gestured left.

The leopard looked at me and blinked its eyes.

I continued with my whisper, 'I've driven all the way from Kora and I'm tired and don't really want to get into a fight.'

The leopard sat down on its haunches and licked its lips.

I took a step backwards and turned to the side, hoping to show that I was in no mood for a fight.

There was a five minute standoff before the leopard calmly rose to its feet, flicked its tail and headed off in the direction that I had asked him to!

Come morning, Kim was adamant that I go back the way I had come.

'There is no way I can take you out in my vehicle,' she said, 'if they catch you, I'll lose my job.'

And so at dawn I crept out of camp and headed back along my now familiar route.

As I reached the *Kopje* I lay out on the rocks absorbing the first rays of the sun and watching the plains game as they came to life on the savannah below.

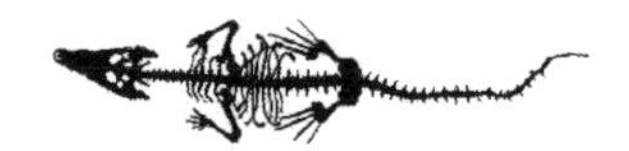

Dominique's face stayed with me as we paddled between the Paradise Islands for several more hours, stopping to look at the massive mature figs and their buttress roots, the ferns, epiphytes and mosses. Parrots called into the emerald depths and hoopoes scurried along curved branches. Kingfishers were everywhere and bright ripples "staccatoed" out from each precise dive, while water-bottle birds "puk-puk-pukked" through the canopy gloom. Crocodiles are a part of paradise too, and there were plenty of them. One monster that lay sunning itself was a good 12ft.

On Imire a year before we had had to catch one of these amphibious lizards to relocate it from the dam it had chosen, to a bigger dam that was off the property and not used by swimming children.

The tale is, in retrospect, comical but at the time we feared for our young, roulette lives.

We were woken early one Sunday by a radio call from Pat. Pat was Reilly's grandma and she was, a feisty old lady of nearly eighty who woke early each morning to go fishing, loading and unloading her little rowing boat all on her own despite our pleading for her health.

'I'm radioing to question your work as game park managers,' she said, the static giving her words anger. 'I've just had a crocodile eat little Molly!' Molly was Pat's dacksi, who like Judy's dacksi went everywhere with her. 'I suggest you come up here now otherwise I'm going to shoot that crocodile dead myself!'

We quickly called Olivey and loaded the Land-Rover. The three of us zipped up to Pat and David's to survey the situation. Our arrival found a furious Pat with a muddy, tear-stained face and a calming Dave.

'I'll not have it, Dave. I simply will not. This is where we live and fish,' Pat was saying to Dave as we arrived, 'and those boys can jolly well get rid of that croc before the day is over!'

It appeared that Pat had spotted the crocodile just as Molly had leapt from the boat to swim to shore. The crocodile had moved in despite Pat's shouts and paddle slaps upon the water. Poor Molly had been taken in the blink of an eye. A crocodile eye.

'It's alright Pat, we'll catch it and put it in Mutemwa Dam, nobody swims there.'

Not one of us had any idea exactly what a crocodile trans-location entailed; there was no option but to try. Olivey suggested that late afternoon-evening would be the best time to try and, not knowing any better, we agreed. The day found us practising with a lasso and loading all of the necessary equipment that we may need into the car: headlight and car battery (our makeshift spotlight), ropes, karabiners and baited croc hooks as well as strong ropes to bind its snout and feet for transport. By five that afternoon we were ready to begin.

Of course by now the word had spread around the farm that we were attempting a croc capture, and a small crowd was there to greet us on the dam wall as we unloaded and prepared. We had a small boat upon which Reilly, Olivey and I balanced with only two paddles for steering.

Round and round we went, lapping the dam as we tried to lasso this crafty crocodile. Our lasso skills left much to be desired and each time we threw a well-aimed noose we would miss. By now darkness was well on its way and after a short break we came back to the job at hand. Still the crowd watched as the spotlight lit a path for us. About eight p.m. I threw a lucky throw and managed to

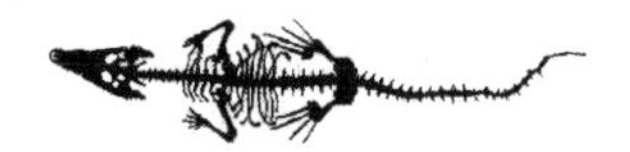

tighten the noose about the nose of the crocodile.

'Here Olivey hold this,' I instructed, 'we've got the bastard! Hold tight while I tie this end of the rope down.'

Reilly was on the paddles and was attempting to hold our location but the crocodile dived deep. We had made one grave mistake: I had tied the rope to the middle of the rowing boat! As the croc dove deep it jerked the rope in Olivey's hands and unbalanced him, with a shriek Olivey went overboard.

'Ahhh,' the crowd said.

Still the croc pulled and twisted and without too much time between splashes both Reilly and I followed Olivey into the drink as the little rowing boat was capsized by the fighting weight that pulled it over to its side.

'Watch out!' the crowd shouted.

'Olivey keep a-hold of that rope,' we gurgled at him as we splashed about trying to right the boat. Our spotlight had been lost in the kerfuffle and now lay at the bottom of the shallow dam, its beam shining up through the water and silhouetting our frightened scrabbles for safety with an eerie glow. Stage lights for the audience.

Olivey held tight to the rope while the crocodile careened about us, snout still tied closed. As Reilly and I clambered back into the boat we heaved Olivey in and quickly re-knotted the rope to the bow.

Now *we* were in control.

'Well done,' the crowd, or was it Judy, shouted.

Eventually the crocodile tired and we were able, all three of us, to drag it in.

At the dam's edge John helped us bind and secure the young croc. Midnight saw us ready to drive to Mutemwa and release the trouble-making

reptilian under the cover of darkness.

'Job done,' Reilly whooped, 'happy farm, happy granny!'

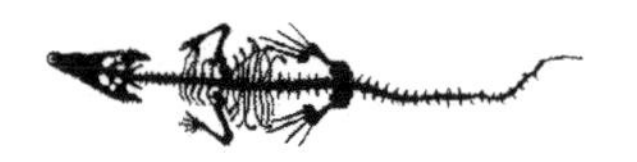

Chapter Nine

Goma – Caprivi – Vic Falls

Thou shalt live in accordance with the rules of nature

As we meandered over the shimmers of the mighty river, tales of crocodiles were rich in our repertoire, and in Goma we stopped for supplies and met an undercover cop called Uncle Joe who said he had arrested a crocodile once! We shared whiskeys and regaled him with every tale of the river. Uncle Joe made us laugh so much that we decided to stay the night. That evening we tried every wily trick in the book to get Uncle Joe to tell us what assignment he was on and how he had come to arrest a crocodile.

'How big was it?' Ace asked.

'He was but a small one. He did not know the rules yet.'

'Rules?'

'Of town.'

We were getting nowhere with this line of questioning. Uncle Joe was determined to be vague.

The night grew shorter, the whiskeys seemed constantly full. Finally as we stumbled towards our hammocks Uncle Joe called after us, 'he was loitering in town. That is against the rules and ...' Uncle Joe stumbled as he zigzagged away from us '... he was trying to look innocent!'

Ace managed half a reply '. . . What do innocent crocodiles ...' and SPLASH he fell into the river - probably within view of another innocent reptilian lizard!

Ace dragged himself cold, dripping wet and

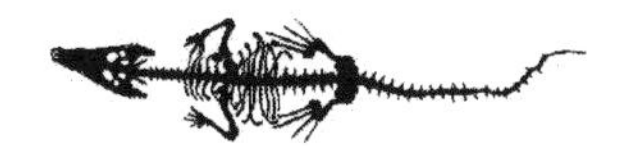

sobering fast from the river. 'No innocent crocodiles for me thank you!' He set to building the fire up and drying off while I poured another whiskey to warm him.

In the morning we groggily fare welled Goma. Beyond Goma lay the Caprivi Strip. The Caprivi is an emerald panhandle of river. Its bulrush marshes spill snow white cotton-wool from velvet brown cocoons. The marsh weavers collect it and built eiderdown nests. Our eyes swept wide from North to South across the habitat of elephant, wild dog, buffalo and zebra. Away from the lush riparian undergrowth lay endless thorn-scrub that rushed boundlessly away to the dry sands of the Kalahari behind us. MMBA is the phrase we use – miles and miles of bloody Africa!

The Caprivi region is rich in minerals and is fast becoming a base for ecotourism. It remains as a vitally important wildlife corridor through the countries of Angola and Zambia in the north and Botswana, Namibia and Zimbabwe in the South. There are many community run conservancies and forests, and here and there we picked out the classic *makuti* thatching of little chalets that belonged to fishing and wildlife lodges.

Back in 1890 Namibia called itself German South-West Africa. The country was harsh and wild and the only access to water was the rugged coastline where the icy cold Benguela current rushed by, bringing with it nesting penguins and wild storms. This lack of water was not to be tolerated amongst the organised Germans, and so a German Chancellor, Leo Von Caprivi, negotiated the acquisition of the land on the Zambezi in exchange for German interest in Zanzibar. This gave Germany an all-important access to the East Coast and to the fresh waters of the Zambezi. But the mighty river never became the easy East Coast access that many

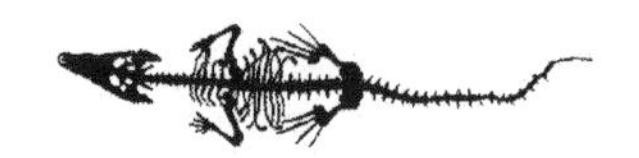

business men dreamed of: Victoria Falls served as a natural barrier to all river traffic.

Fresh-running water will always attract disputes, as has been proven all through history, and in the late 20th Century Namibia and Botswana disputed the southern boundary in the International Court of Justice, fighting mainly over a large island that lies in the Chobe river, a tributary of the Zambezi. Even now, politics in this region are unstable; but it was hard for us to imagine any skirmishes here as the wide, slow flow snaked between the creamy sandbanks that were rippled with the imprints of the crocodile's tail. We simply could not imagine battles being fought amongst the chorus of the kingfisher and the sandpipers.

Once, at dusk, we came upon a small houseboat where the clink of ice-filled glasses created a mournful duet with the hippo as he sang into the night.

'Pss-ahhh!' he said ecstatically and climbed out to stand on a sandbank in-front of the setting sun, twitching his ears and preparing to trek inland for grass. These were the noises of each evening as the flamingo-pink dawn and dusk sun shone down across the swirling waters.

But this was the dry season and the climate would soon change, bringing in storms of both mosquitoes and rain, and soon it would no longer be bliss to lie out under the stars as we had been doing. The sand-bars would lie three foot under until the next dry season.

Late one lazy morning as we were about to stop for lunch and a siesta, we met a senior-chief walking along one of the river banks with his escort of lesser chiefs. They appeared to be discussing senior chief matters. He stopped us and we introduced ourselves. We told him where we were going and what we were doing. He was impressed

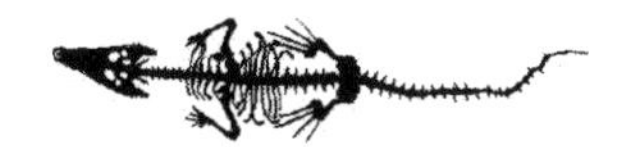

and advised us to meet the Prince of Barotse who was currently residing in the region of Mwanda, a little downstream, as the prince too had a kayak and would be pleased to receive us.

Now, if there is one thing in Africa that will befuddle you to the core, it is the local African's ability to give directions. Most rural inhabitants have never seen a map or been taught how we western educated people direct. Here we were, in amongst an ever-changing river being directed to one specific spot on the river bank.

'You pass down this channel and move to the left to the next channel, where you follow straight till you turn and then you follow ...,' the instructions were long-winded, difficult and frankly nonsensical.

We bade our farewells and headed off to find the Prince's home. Why not, we thought. Who could resist meeting a prince?

'This way' Rei called.

'No definitely this way,' Ace countered.

'The middle channel is a good bet,' I added.

We each won and finally after hours of fruitless searching, and on the verge of giving up, Ace spotted an old, worn kayak tied to a tree. We pulled in, moored our kayaks close by and climbed the mud steps that had been dug into the bank.

A young girl met us 'Good day,' she said.

'We are here to meet the prince,' we chorused.

'Yes he will be pleased to meet you. I will call him now. Please be patient with him.'

The prince welcomed us with great ceremony, bowing and smiling and holding our hands, all three at once.

'I am the Prince of Barotse,' he said to each of us in turn holding all our six hands in his two. '

We stood nodding, smiling and making small talk in an overgrown garden on an island on the

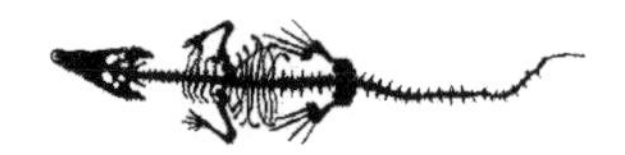

Zambezi, hand in hand with a prince! He bowed and smiled again and we bowed and smiled back.

'Perhaps you would like to see the tourist lodge I am building?' he asked with a shining face. Then his face fell hurriedly, 'but you cannot stay there. It is as yet unfinished.'

We followed him around his lodge which he had obviously had grand plans for and had abandoned half way. Huge stacks of cement had hardened through; orb spiders had made webs between untreated beams and birds nested in the half thatched roof. He seemed not to notice and smiled as he led us from room to room.

'Come, now I will show you my house,' he proudly said. 'I am a modern man and have many appliances.'

Appliances indeed. There were TV's all over his house. It was a small house, built by a builder who knew nothing of straight lines. Doors were mal-aligned and windows so dusty you could barely see out.

'But this one is *my bedroom*,' the Prince took Ace's hand and we followed. 'Look,' he said puffing out his chest slightly. 'I am sleeping like a real Prince!'

Instead of a bed he had a stack of ten mattresses facing two TV's both featuring a bad picture. It was a wonder that a fairy-tale had made it this far up the Zambezi, if indeed it had!

Back in the overgrown garden we rested on old, smooth worn wooden benches and exchanged tales until the Prince told us that he must go sleep.

'I will see you tomorrow; you must put your tents here to be comfortable.' He gestured to a very small patch of saw like grass.

'Not a chance I'm kipping there' Rei smiled after the Prince had left. 'Those riverbank trees over there look *lekker*.'

The maid appeared with a smile and told us that the Prince coming outside with us was the first time he had left the house in months, in fact the whole of the dry season. She said he was sore of the heart but could not and did not want to say why.

We spent a quiet night discussing princes, mattresses and aching hearts. Lying back, arms stretched up above my head I thought about nobility. I thought of the princess who lay buried beneath Imire's Castle Kopje in Zim shrouded in gold bracelets and beads. Her lineage is believed to date back to the 1600's and, as per the custom of royal burials of the time, she stands vertical beneath the stone, ever a listener to our sundowner tales. Perhaps she too slept on many 'mattresses' or grass filled mats.

In the morning the Prince came down to look at our kayaks and loved them but seemed gladdened when we loaded up, bade our farewells and paddled into the slow moving current.

The morning's paddle brought us to a set of rapids just up-stream of Katima Mulilo from which several horror stories sprung from past paddlers. Back in the old days, the Lozi paddlers, a dozen to a boat would navigate these rapids only if the water level was favourable. But they knew and understood the water-levels.

We, on the other hand, had no idea.

We pulled our boats up onto the sand bank and climbed down amongst the boulders that littered the sand, so as to scout out the line. We knew that back at Sioma Ngonye the water-levels had been more than favourable and so we guessed the situation would be similar here. Reilly drew the short straw.

'Right boys, follow me,' he grinned as he kitted up, making sure the teapot was tied down securely this time. Ace's hands shook slightly on his

paddle as he took a firm hold and I too had adrenalin coursing through my veins. We simply couldn't chance going over.

We had yet to learn that crux manoeuvre. The Eskimo roll that is perfected with a paddle sweep and a hip flick. Practice had proved that although we had perfected the first half of the manoeuvre – the rolling over so as our body lay trapped in our kayak underwater – we were nowhere near perfect in our rolling up!

From the bank we could see one nasty recirculator.

Recirculators occur when the accelerating water comes upon a sudden dip and thus the water 'digs a hole' for itself, the water immediately downstream falling into this hole while the main flow continues underneath. In kayaking terms (so we were told!) a recirculator is best avoided as it has great capacity to trap canoeists where it will spin them around for a while before spitting them out downstream (a bit like Roald Dahl's stone-chucking, snozwangler). If the recirculator happens to be a 'keeper,' well then that is trouble as the keeper will never throw the canoeist far enough into the down-flow for him to paddle an escape, instead he will be sucked back into the vortex, churned over and over and over. A demise we did not wish for ourselves. We paddled and braced, and braced and paddled. All of our kit was once again sodden but we were through and safe and as the current swirled us downriver we whooped and cheered.

'*Yakanaka! Yebo*! *Yeehah*!' we yelled joyously.

In Katima Mulilo we stopped to re-stock our dwindling food supplies but only quickly as Vic Falls and good food beckoned us to hurry. We still had to navigate the smaller Mambova and Katombora rapids but this we did without mishap or comic

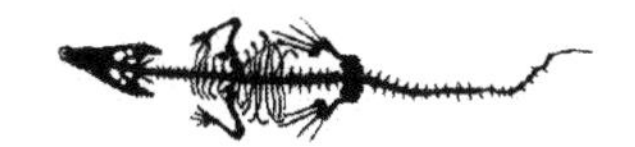

incident as we scouted them well and planned our route down carefully.

To break in with tales of rapid running I recount to you a very recent set of rapids that Lovat (my *rafiki* from Kenya) ran blind and nearly didn't make!

Just last year Lovat and I had decided to run a ten kilometre section of the Pekera River in the Rift Valley in Kenya. We left Naivasha at dawn as the mists were clearing before the powerful sun. Our destination was a region where flash floods had moulded landscapes and features that were more reminiscent of deep canyon country. Thorn trees and the introduced *Prosopis* were rife and could rip deep into your skin with no warning.

We parked the little Suzuki on the 'Old Bailey' bridge and looked down at the chocolate water that ran sedately beneath.

This section of the Pekera I knew. As children we had held birthday parties here, happy groups of six year olds that were engrossed in sugar and intent on making a name in mud-fighting, while parents floated in old car inner-tubes keeping a watchful eye on their brood, in case a sly crocodile should creep up. Beyond here the river was unknown; we had zoomed in as far as was possible on Google Earth and, as we unloaded the large expedition kayak and the smaller play-boat we could not have known what we were about to let ourselves in for.

Lovat had barely kayaked in his life and neither of us had any experience in the river-running of technical waters. We slid our kayaks down the bank and into the water over the smooth, rounded pebbles that made up the Pekera.

'Meet us at Marigat Bridge in a few hours,' we told another great friend of ours, Ndung'u. With that we snapped our spray decks over our cockpits,

tightened our helmet straps and began a war of attrition. For a couple of kilometres the water ran smooth and calm, Lovat lifted his paddle over his head and whooped.

'*Yeee-ha* Jamie, this is the life!' his voice echoed back down the loose rocky walls of the steepening bank and the hyrax scuttled away in fright. Soon the pace of the water quickened and began to test our navigation. Lovat, in the larger expedition craft, wasn't too hot at keeping the kayak in a straight line and would T-bone boulder after boulder. I side scraped many a boulder too leaving yellow kayak paint on the rough river rocks to mark my progress.

But now the gradient began to fall at a constant rate and our heart rates quickened in anticipation.

'Don't go left Lovat! Keep right!'

'Watch that hole!'

Soon the trouble began. I was the first to swim – an easy feat when I couldn't roll up after I was flipped by a nasty log that shunted me left into a massive boulder.

Lovat flipped behind me and paddles, boats and bodies collided. Lovat was heaving for breath and I'd lost a hold on my kayak. We scrambled onto a boulder and saw it heading around the corner, swept along by a current that had carved deep into the bedrock over millennia.

In the kayak was the GPS the mobile phone and our go-pro camera!

'Jamie, go, go, go!' Lovat shrieked as I dived into the flow and followed my kayak, knees scraping the shallow bed and elbows colliding painfully with trapped flotsam. Soon I had caught up and heaved the full kayak onto the bank to empty it. Lovat paddled in to help.

'Shit, Jamie, looks like the river's not getting

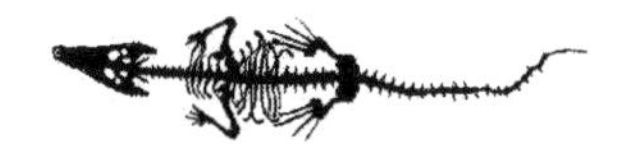

any calmer. Look ahead.'

Ahead the cliffs began to stretch taller and the gap between the walls narrowed.

'Once we go in there, Jamie, we ain't coming out, look how high those cliffs are.' I squinted at the blue cracks of sky between the scaffolds.

'It will be fine.'

And so we snapped in again and the gorge walls closed in on us. Green figs clung to ochre fissures that were crumbling under pressure of the roots. In one place the river was forced through a narrow opening of only a few feet that was jammed with boulders. We worried that we could be caught in a rock sieve and so had to portage section after section of these potentially lethal boulder gardens. Again and again one or another of us swam and, as our skills of rescue were few, it was self-rescue each time as our boats were swept around corners and down frantic currents of swirling wave crests.

On one corner we could hear a deep rumble punctuated by sporadic and resonating knocks that we couldn't place. We pulled our crafts up and boulder hopped around the corner. Here the water was separated into fast flowing channels that were forced through narrow openings below which deep potholes had formed. It was from these deep potholes that the knocks came, we could hear the trapped rocks rolling with the swirling water, eroding themselves into perfectly smooth and ever smaller boulders, then stones, then pebbles; the water muting the sound, as they knocked against their rocky prison wall.

Portaging was challenging and we risked sprained or even broken ankles on the smooth boulders over which we hiked, kayaks on our shoulders, then balanced between us, and then dragged. Our drinking water now was finished and the couple of apples had not been sustaining, but

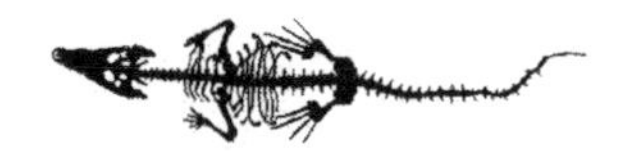

still the rapids had not stopped. Soon we heard a deep roar and in front we found an enormous waterfall that would have been lethal had we have run it. But we portaged, and at the bottom pulled out the GPS to check our mileage. We had completed only three kilometres of the ten!

'And at least two of those we have either been swimming or walking!' Lovat exclaimed. But there was no way out and so we continued.

And then I had a nasty swim and Lovat a nasty surprise.

As I was caught in a whirlpool, my little, totally inadequate play-boat was swept over and I was roughly held against a rock. Another force of water swept the boat round and my helmeted head connected hard with another rock. I pulled my deck and gasped for air as the stars cleared.

'Lovat grab my boat – quick!' But Lovat was too slow and my boat was caught in the fast current. It disappeared, sucked under the muddy brown water as I went after it but the current was violent and my toe was torn back, broken. My shoulders and hips connected painfully with rocks in my path. Finally, a good five hundred metres later, I could see my boat no longer.

'Lovat man, it's gone.'

'No, no. I saw it go down and I didn't see it come up again,' Lovat panted as he caught up with me. At that moment the boat, half drowned, floated softly by.

We both sighed heavily. Wearily. I snagged it and once again emptied it.

'Lovat, this is getting serious. We gotta get out of here – we are going to kill ourselves if we don't.'

'Yeah but how, Jamie? Look at those cliffs.'

'Look there is a spur coming in a few kilometres down, maybe there is some track up the

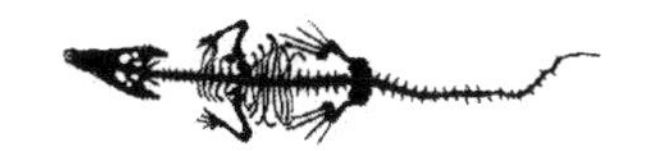

spur.'

We climbed wearily back into our kayaks and managed another two hundred metres before we again had to portage. As Lovat lowered his three metre expedition craft down into a pool, a frightened crocodile ricocheted out of the water and up towards him.

In his fright Lovat lost his balance and slipped into the pool!

Were there more crocodiles? He didn't hang around to find out!

A half hour later we came upon a lone honey hunter whose bee hives hung from emerald figs and who himself hung from intricately knotted ropes with a smoking handful of greenery. This he planned to use as a bee sedative so as to collect their honey.

'*Mzee!*' we called up to him, 'we need your help.'

He looked down at us through the smoke and slowly began to lower himself to our level.

'Where have you come from?' he asked

'The river, from near Emening there, the bridge.'

The honey hunter raised his eyebrows and frowned quickly. 'But you are stopping here – not continuing? Beyond here the river is very dangerous. Not passable.'

'But how can we get out?' Lovat asked.

His blackened finger pointed up to the spur, 'there is a path there but it is very steep.' He looked across at our boats. 'I am not sure how those ones will be carried.'

We left him to his honey gathering and snaked down to where a steep, eroding dirt path reached down to us from the cliffs above.

'Salvation,' Lovat muttered.

And then the hard work began. Hoisting, heaving, pulling, winching and shoving each other

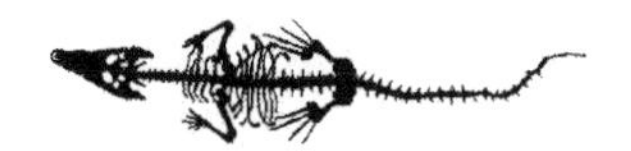

and our kayaks up this almost vertical path that crumbled as we clung to it. My toe was hanging loose and we had had to strap it to the others with insulation tape. Now we needed our toes to grip to the exposed tree roots that gave us traction on the climb up.

Darkness fell as we reached the midpoint of our climb. Aching muscles and painful bruises patterned our torsos and our throats were parched with the exertion. The final hour was in darkness and thorns reached out from the shadows to tear our skin and shred our shorts as we slipped and skidded to the top where the road ran.

We rang Ndung'u and soon he appeared around the bend. Never had we been so happy to see him and we collapsed in such exhaustion that we worried him.

'Let's go to the hot springs at Bogoria, there is no way we can make it home tonight.'

At the hot springs we rested our aching bodies in the hot pool as the stars flickered overhead.

'Let's stay here,' Lovat suggested, 'get a tent and then we can stay in the pool all night if we want to.' Ndung'u checked us in and called us to eat ... that was the normalcy before pandemonium broke loose. On the floor of our tent Ndung'u leapt clear of a small brown snake that Lovat insisted was only a brown house snake.

'Here, let me catch it.' He nimbly grasped the snake at the neck between thumb and forefinger and carried it outside and across the lawn to the reception.

'Lovat, you sure that's not a rhombic night adder?' I called as I ran after him.

'You better check all the tents Madam, we found this in our tent,' he began telling the receptionist when suddenly the snake twisted in his

grip, loosened itself and curled round to latch onto Lovat's forearm.

'Yikes!' Lovat yelled as his reflexes bade him pull violently backwards. But the snake was attached. We pulled back the mouth to see two fangs deeply embedded in a rapidly bruising grip. The blood was beginning to ooze and Lovat was beginning to pale. The night guard took one look at the developing situation and briskly fled.

'*Woi woi woi woi,*' he shouted as he ran, and left us extracting the fangs of an unidentified snake from a woozy Lovat. In minutes he was back with a 'snake-stone,' a local cure for snakebites. He pressed the soft, black stone to the wound and rubbed small circles with its smooth surface. Colour began to return to Lovat's face and his breathing became less laboured.

'Wowser Jamie, we gotta be more careful man. I could be dead.'

'Wait another hour and you might be,' I shot back at him, more worried that I was letting on.

'I can feel this stone working, it's incredible! It feels like something is being drawn out - like I'm being injected with energy.'

And so we both lived to tell another tale, lived through another set of mis-adventures!

Back on the Zambezi the first baobab rose out to greet us at Katombora, its grey-granite trunk almost twenty feet thick and as we drew close to Kazungula we discussed our options. Kazungula was the border post between Zambia and Zimbabwe, and the last thing we wanted was to be pulled in by some official here. We decided that we had no option but to paddle straight through, relaxed and easy. If we were signalled in we would go to shore, but if not we would paddle right on by. I'm not sure anybody took any notice of us at all, and before long

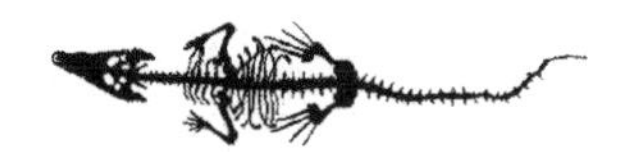

we were out of sight of the border post and on our way to pick up Kate.

Before the trip, Ace had come up with an idea that, had we acted upon it, would have worked well to help in raising money for the rhinos. The idea was to auction off places in the expedition, so as whoever felt they could raise money would in return join us for short sections as we paddled down the river. The logistics of this proved to be too horribly complicated. It would have meant taking an extra kayak or organising meeting points on the river where there was vehicle access. As we were not renowned for keeping time, this was not an option that would work without a vast amount of effort.

However, there was one section where this could and did work. And Ace had organised it. Just below the Katombora rapids we were met by a *rafiki* of Ace's who had raised a seriously decent amount of money. The only other kayak that we had available to us was one that we had borrowed from a mate in Vic Falls. It was a faded blue and sported a large hole near the end. It was dropped off with our esteemed fund-raiser, Kate O'Donoghue, just past Kazungula.

Reilly and Ace both offered to swap their kayaks for this 'German Sub' as we nick-named it, and Kate was very wise in accepting.

Kate was a shortish, tanned girl who smiled a lot and who delivered us biltong, steak and other coveted treats. I remember the first of the three nights that we spent with her on the river. Reilly served bile-free crocodile tail on a bed of rice for dinner. This for us had become normal fare and we didn't think to ask a newcomer if it appealed! I think that perhaps Kate had brought some of her own food along which she ate quietly out of our sight at night, for she certainly didn't eat much at all, or perhaps in hindsight, she could have been

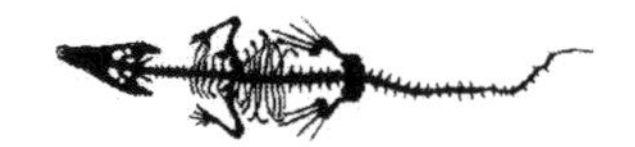

vegetarian!

I don't blame her not wanting to share her snacks with three ravenous boys.

Company did us good and drew our conversations away from fishing and paddling. Our jokes improved and our language became more refined. Nature's courting instincts kicked in and each of us tried to be more gentlemanly than the other.

Having someone along did show us what a tight bond we had formed and gave us a new perspective on how much we had experienced.

The wilds crept quickly back after the blotch of civilisation at Kazungula. Now we saw elephant and water-buck, kudu, warthogs and buffalo on the river banks. There were still a few villages on the Zambian side; messy affairs of sticks and mud. The few fishermen that lived here fished at night in dug-outs with lanterns tied to the front. One night we persuaded two of these athletic, smiling fishermen to take us out with them. As they prepared, we told them of our journey but the looks of disbelief on their faces showed that for them this was their world, this small section of the Zambezi. What lay far above, in the regions where the water flowed from, was of little interest to them.

When the dug-out was suitably prepared, we three boys climbed in. One fisherman sat at the stern working his paddle in small silent circles that propelled us ever so gently forward. The other man stood poised, with a four pronged trident-like spear, at the bow. We pushed into the marshlands where the bull-rushes grew tall. Now and then a night-heron would lift like a shadow into the blackness, or a roosting bird, disturbed, would utter a startled call. *Vundu* or catfish, would curl to the surface to feed on the small fry that were circling within the reaches of the lantern. When the moment was right the man

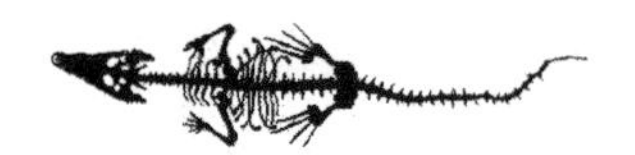

on the bow would launch his trident-like spear up to five metres and the *vundu* would sink into the water's depths. On every throw they would have speared two or even three of these tasty catfish, their ability and aim so exact. We tried too, but it took us numerous throws to spear even one of the *vundu*.

Zambezi Bush Kebabs

Kebabs add an essence of architecture to any meal. This can be useful on any long expedition where meal repetitiveness can lead to loss of appetite.

Mixed kebabs with nuts, fruit and meat add all the required nutrients to the body; the proteins, the fats and the vitamins.

Kebabs are always delicious. Nuts embedded into the meat, spices, herbs and fruit add to the full zest of flavour but here on the river we have to cook with what we have. Always keep your eyes peeled along the river for different edible fruits.

Serves 3 imaginatively hungry kayakers

3 x long green sticks
1 x 2kg tiger-fish or tilapia
1kg crocodile or monitor lizard meat
Wild figs
Sugar or honey
Chilli (oil) and/or salt

1. Collect a good stock of wild figs and sugar them liberally. If you have collected fresh honey then rather use this over sugar.
2. Cube all of the meat that you have; crocodile or

monitor, tiger or tilapia.
3. Alternate the meats with the figs on the skewer.
4. Cook over hot coals until nicely browned.
5. Use chilli and salt as needed.

There were many islands and channels and the moon was nearly full now. The swamps seemed swathed in a tranquil light each night which inspired stories around the fire. Here there was no need for ghost stories. We only had to talk about the crocodile and hippo attacks that we had heard tell of to scare both ourselves and Kate.

'Gustave,' we asked her, have you heard of him?'

Kate shook her head and we filled the darkness hours with tales that left us all tossing and turning in our hammocks. Gustave was an enormous crocodile of over three metres who had at least eighty victims to his name and a television show. But he also had a protector, Patrice, a man who had been brought out to Burundi in 1998 to hunt down and kill this slayer of men, women and children. But the cunning Gustave eluded Patrice as he roamed his large territory from the Tanganyikan lake shore to the mouth of the Rusizi River, up the river some sixty kilometres and across the fields that bordered the river. Patrice had grudgingly formed a great respect for this gigantic Nile crocodile whose large head was scarred with old bullet wounds and whose name was whispered fearfully at night to naughty children.

'Where people cannot respect the territory of a crocodile, then they will be eaten,' we told Kate as she eyed the water before us for any lurking beasts.

'And Gustave is supposed to weigh one tonne!' Reilly exclaimed

'And is seven metres not three!' Ace added 'that's what I read somewhere.'

'I tell you, a croc like that has the strongest genes of any croc left in Africa,' I said and we quietly mused in the darkness.

'I mean what would you really do if you were taken by a croc?' Kate asked.

'Well ... I guess it depends on the situation, but here, now, we are as ready as we can be,' Rei said. 'We each have a knife on our belts and one strapped to the side of our kayaks – I would grab my knife –' Rei leapt to his feet and demonstrated pulling the knife from its sheath, '- and I would slam that knife point into the croc's eye if I could!'

'Or the soft skin under its throat.'

'Really?' Kate doubted us.

'Well it's live or die,' Ace said, 'and personally I'd rather take the living chance.'

'And plus,' I turned to Kate, 'we have these little nine ml pencil pistols that we have made with a trigger switch that will explode underwater and hopefully create enough noise to startle the croc into letting its grip go so we can get away.'

'And if it pulls you under you play dead,' Ace instructed, 'let it think you are already dead so it doesn't roll you and then if it gives an inch you take a mile.'

Kate doubted that she would be able to do any of these things.

'Don't worry, we'll protect you,' I said, 'unless we meet a Gustave – then my friend, you are on your own!'

Just on dawn, guinea fowl would wake us with their coarse calls, while in the afternoon kudu, oryx, eland and sable came down to drink at the water's edge. Some days it seemed as though we were in a fairy tale as the fairly open thorn scrub spat forth the occasional monstrous baobab and many a startled antelope which would raise its head and watch before lightly plunging back into the

undergrowth. Brown, sturdy water-buck broke into rugged charges through the scrub or froze, their velvet muzzles to the wind.

Lodges began to appear on the river's edge but the wildlife did not stop. Zebra turned their fat rumps at us, buffalo gazed demurely towards us, and elephant lumbered grey and light along the sandbanks. African Skimmers filled whole banks with their pied colours and red beaks. The islands were verdant wildernesses that the elephants swam across to each afternoon. The spoor on the river banks told of night prowling leopard and hyena, and we found the scats of genet and serval, tortoise and aardvark. Now and then we saw little families of warthog wallowing in the shallows. As they picked up our scent or heard us, they rushed to their feet and stared up-river, tails pointing up to the heavens. Every time we paddled past islands, large or small, wildlife would come crashing out of the brush on the other side. It was a rugged beauty, topped with big flocks of vultures and raptors of every sort.

These days in Africa, big flocks of vultures are a rare pleasure to see. Vultures are amongst nature's most successful scavengers and are perfectly adapted to this important lifestyle. Now over sixty per cent of vulture species worldwide are threatened with extinction, and Africa is a prime area of concern. The reasons for the population decline are varied and sometimes location specific, but the deliberate poisoning of carnivores by humans is a major factor. They wish to poison the carnivores mostly to stop predation on livestock, but the efficient scavenging vultures will hone in on a kill from miles away and are not able to detect if the meat is poisoned. The ecological consequences of the declining populations are huge. The scavenging of carcasses by vultures promotes the flow of energy through food webs, and their flocking helps other

predators to locate these meat scraps. Vulture numbers have been recorded to correlate in a visible pattern with feral dog species too.

In the Kazungula - Vic Falls section alone, we reckoned that we passed over a thousand hippos. Each pod held thirty or more, ear-flicking, honking individuals. Hippos come out of the water at night to feed and sometimes we had no choice but to travel at night to avoid them; other times one of us would run along the bank so as to scare them to the opposite side of the river, giving us a clear channel to paddle down.

We had been told that old Kingsley Holgate, a legend to all young 'explorers,' was staying in a lodge on the Zambian side and so we made plans to call in and to meet him. But somehow or other we lost track of each other as we formed our own paddling paces related to our interests and we did not manage to find each other again till nine that night.

As Reilly, Kate and I waited behind a glowing fire on a sand bank for Ace, a big shadow cruised by in front of us.

'Ace, Ace, Ace?' Reilly jumped up and shouted, each call louder than the last.

He sat down again, '*that* is a big croc.'

Minutes later Ace pulled in and we all breathed sighs of relief.

Against the advice of many, we had no rules for re-grouping or paddling together. Our interest in different things meant that we paddled at different rates. It was a system that worked well for us. All of us were gregarious loners who were content in our own company as well as in the company of others. In hindsight we realise how utterly dangerous this was and how lucky we all were.

Through all of our lives so far, we had mixed with nature and people fluently. Now the peopled

town of Vic Falls called to us and the next evening our hearts lifted and our muscles lightened as the lights of Vic Falls twinkled into sight. Late at night we passed Prince Christian and Princess Victoria islands and pulled in, ready for some R&R at 'The Big Tree.'

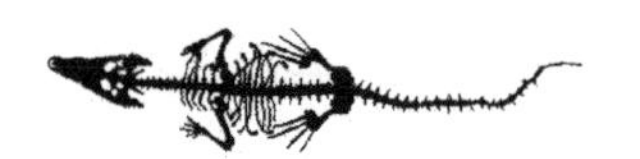

Part Four

The Melody of the Middle Meanders

The secret is how to live. Only it is not a secret at all, but the balancing of respect, good luck and karma.

Nothing breaks the monotony of the thorn scrub except the occasional explosive moment when a large baobab breaks surface like a whale rising from the ocean.

Thomas Pakenham, The Remarkable Baobab.

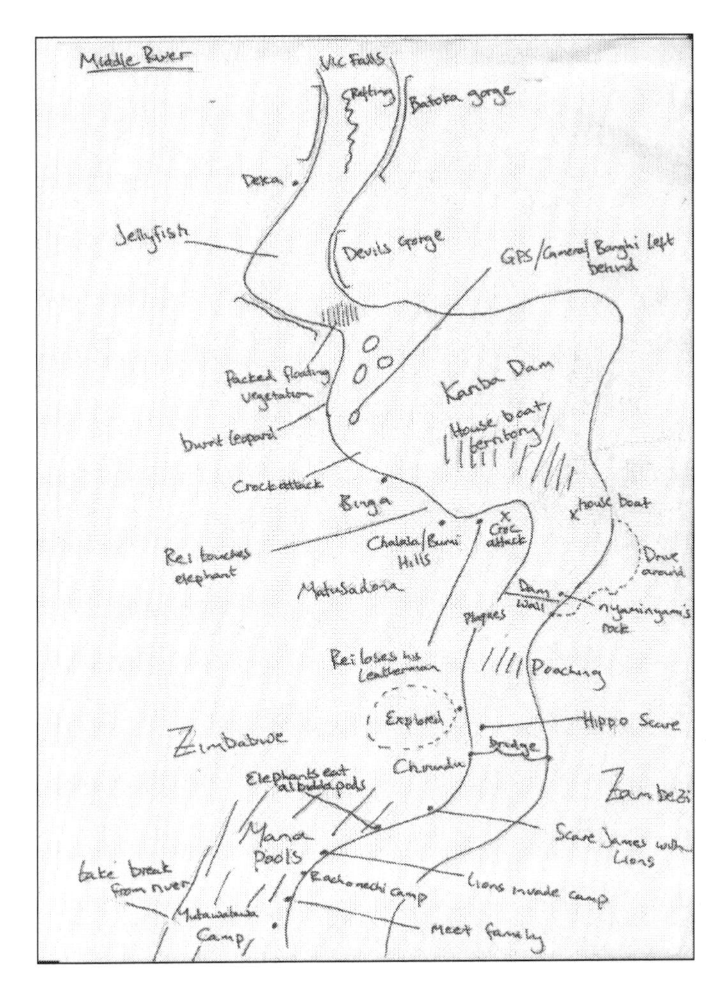

Sketch map of the Middle Zambezi

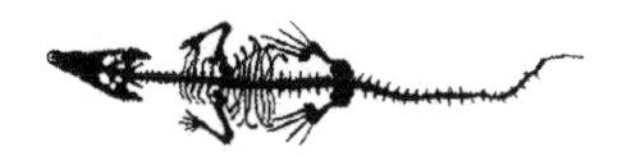

Chapter Ten

Vic Falls – Devil's Gorge – Kariba Dam

Thou shalt respect all of Africa

Victoria Falls town is a town like no other. It is not advisable to go wandering about at night as there is a very real risk that you may walk into an elephant, or a predator. In fact on one of my stays there I did just that. Having just been at a friend's house for dinner and a couple of beers, Reilly and I made moves to head back to our accommodation. We had arrived just after lunch on bicycles and now hopped on to pedal home. Of course we had not planned to cycle back after dark and so only had a small phone torch for light (the street lights in Vic Falls not being very reliable!).

The night was calm and still, the streets were deserted and we had to ride slowly so as not to ride into the potholes and uneven curbs and drains that makes part of the Vic Falls infrastructure. Not far from home we heard the unmistakable sound of a shaking tree. We stopped. The noise came again. A shaking tree meant only one thing – a feeding elephant.

Reilly crept ahead and after a few minutes came back shaking his head.

'Don't worry all's clear.' He said.

As we mounted again the elephant trumpeted.

'*Whoooo-eee*' it shrieked and we shook at the

knees, ditched our bicycles and dived into the bushes. We sat quietly, still shaking with the unexpected night disturbance. With no fanfare a large elephant lumbered out of the scrub on the other side of the road and headed the way we had just come from.

This was a typical night in Vic Falls and the daytime was supplemented with warthogs and vervet monkeys, elephant, troops of banded mongooses and sometimes, on a dark night with no moon, even buffalo! They wander through the town, on the paved roads like any tourist, except they are the real residents, the real 'zebra crossings.' And if you get caught in a traffic jam it is very likely to be caused by an elephant on the road reaching up with his long nimble trunk to pull some especially tasty-looking foliage down.

Everywhere there are large baobabs, slate grey and as old as the hills. If trees could talk they would tell of Livingstone and of Shaka Zulu, but they remain coy and sullen, giving away none of their secrets.

We stayed for a week in this somewhat old-time town that buzzed with adrenalin craving tourists. Both Reilly and Ace had organised for their girlfriends to come down and meet them. This allowed me time to myself.

In my diary I've written '*Vic Falls I love you*' and that simple line still rings true today. Despite an old torn sign in one of the bars that reads: '*We regret to inform all customers that the light at the end of the tunnel has been turned off until further notice*', there was still a strong pulse of life.

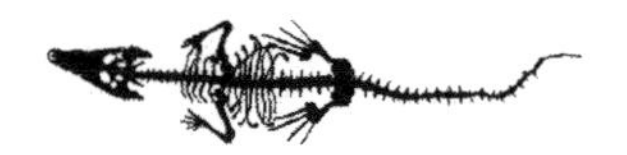

Elsewhere in Zimbabwe tourism was at an all time low. Politics, inflation and violence had robbed people of the desire to travel inland, but here at Vic Falls people surged everywhere, ebbing to and fro in the bars as they swapped stories and joked. I too got caught in these sudden crowds, but sometimes I ached for the less superficial, and broke away and visited the old-time friends I had made there, and walked to the places that were iconic in my mind. I took many a long walk to see the thundering falls and to muse in the damp spray that enveloped them. I crossed the bridge into Zambia and visited mates in Livingstone, I took time to go and pay homage to old David Livingstone's statue and I caught up with my old Africa *rafiki*, Francis Mudavadi whose great, great grandfather had met Livingstone. Mudavadi's wrinkled face and white-toothed smile could bring a sense of fullness to anyone. I especially loved hearing his stories of the old times; the old people, the wildlife and the legends that are a part of all African cultures.

Like all the old men from the Batoka tribe, Francis had a '*sansa,*' a traditional instrument, that he loved to play far into the night. He sang many a song of battle and drought but in his words too he spoke of Livingstone; of his long caravan of porters and of their tales of the land through which they had passed.

One story that must have captured his young mind as his passion in the telling was rich, told of the fleeing Batoka from rich lands around the falls where their cows gave more milk than could be used. The Batoka were driven out of their lands by the invasions of the Moselekatsé and Sebetuané. At the same time, other fleeing tribes from Bechuana and Basutu did not know how to swim and when they sought passage across the Zambezi they were

betrayed by a Botoka chief who ferried men and women to separate islands. There leaving the men to starve and appropriating women for his people. As he recounted this tale, mournful notes escaped his *sansa* and drifted on the air with the water spray from the great falls.

It was here, facing all these jumbled pieces of history and beside one of the worlds natural wonders that I realised that Africa is a mode of life. Here we live our lives in a landscape of greatness that sometimes can feel so very great that it is heavy on our shoulders. Everywhere one can find awe, whether in the quiet of a vast savannah full of waving grasses or beside this, the highest waterfall in Africa. This is an awe that dwarfs any man, stirs any heart and lightens any mood. There are few that are not able respond to this sheer splendour with smiles and a lightened heart. But every explorer must remember that exploration asks for inner solitude, a precious gift one has to work hard for.

Some days the three of us met up. One day we paddled for hours in the pools at the top of the falls. There is a pool at the very top of Vic Falls that seems to put you on the absolute edge of the hundred metre drop. Here we lay for hours beneath the lemon yellow sun as the water plummeted, sending up continuous spray from a frothing pool hundreds of feet below, where the water then thundered on its way into rapids that were swift and white. Here we heard the story of Matt Blue and the hippo.

Matt Blue was a raft guide and kayaker who had lived and worked on the Zam for years. During one particular week they found a dead hippo washed into an eddy. Day in, day out this carcass remained

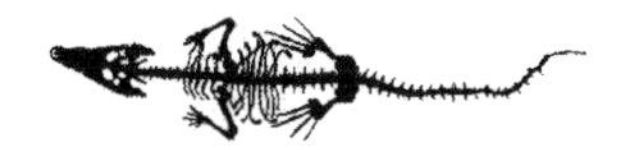

trapped in the eddy, rotting slowly. Soon the stench became too bad and the guides were asked to remove the carcass so that clients did not have to endure its sweet fermentation. Matt and his team pulled into the eddy with their raft.

'How are we going to pull this monstrous river horse out into the flow?' they asked. One local guide had an idea; he would climb onto the bloated body and hold tight onto one end of a rope that would be tied by its other end to the raft. The four others would then paddle hard into the flow, dragging behind them the hippo and its 'rider.'

The boys agreed; after all this plan seemed sensible.

The four in the raft took their positions and made ready to paddle, while 'Steve' took his end of the rope and climbed onto the carcass. But the hippo was so rotten 'Steve' fell through its rotted flesh and into its bloated stomach cavity – the rafters hooted with laughter and paddled hard towing 'Steve' and his steed into the flow. 'Steve' hung in and managed to ride two full sets of rapids with only his head and shoulders showing as the hippo tipped and swung wildly in the white water.

We spent another of our days game driving in the Zambezi National Park through *mopane* bush land, twisted gnarled trunks on basaltic lavas over two hundred million years old. There were teak trees and golden grasslands that hid many a shy beast. On one of our drives we rounded a corner and came face to face with a surprised bull elephant that snorted heavily at us.

'*Whoosh!*' he said as he expelled air heavily through his long grey trunk. His great convex head

nodded and his two ears flapped like sails on a Zanzibar dhow. He reared up on his hind legs and flapped those enormous ears trumpeting 'wheee!' as he thundered past us and into the dry thorn scrub and mopane in a flurry of red dust.

This elephant charge reminded me of Colin Waddle, a great friend of Reilly, Ace and I, who had been forced to leave his job because of a charging elephant. At the time he and his wife Nan were offered the position, they had not been in work for a while. The lodge they were to manage lay in the Zambezi Valley beside a hunting concession. They arrived with a few days to spare before their first guests were to arrive; they filled their days with learning the roads around the conservancy, learning the runnings of the lodge and getting to know the staff. One evening a large bull elephant wandered into camp.

'That is Moozie,' the local staff said, 'he comes here every year when the hunting season starts. Guess he knows that here in our conservancy he is safe.'

Colin moved forwards to get a better look and as he made eye contact with the large bull it froze. And then it took a step closer and tilted its head, flapping its large ears and gazing intently in Colin's direction. It seemed to be thinking and the atmosphere was charged with its stillness. Suddenly it raised its trunk high over its head and as it trumpeted it laid its ears flat against its head and charged. Straight at Colin.

Colin leapt from the small balcony and ran to a guest cottage behind the mess area. The elephant followed close behind.

It is not normal behaviour for an elephant to pursue a chase to the bitter end, but this big bull was not giving up. He began to break branches from trees beside the cottage and to scuff the dirt with his feet.

'I thought I was done for,' Colin told us, 'I mean that guy was not giving up. He had set his eye on me and that was it!'

Colin thought quickly and lured the elephant over to the bathroom side, being careful to quietly open all connecting doors and the main door as he went. Once he had the irate elephant here he ran for his life – through the bedroom, out onto the verandah, down the steps and back to the mess area. But the elephant was hot on his heels. And soon the destruction of the gardens by the mess area began as the elephant tried even harder to get at Colin.

'Eventually I just had to leave,' Colin said. 'I managed to get into the Landcruiser and I tell you we had to floor the accelerator to make it away from that elephant!'

Colin attempted to come home several times in the next few days but the elephant was determined to keep him away. Finally an exasperated Colin drove next door to the hunting concession to ask for advice, he had never before seen an elephant so determined to kill. At the hunting lodge they commiserated but could offer no insight although they mused that a couple of seasons earlier a client had wounded a bull elephant. Perhaps this was the one. Colin was furious that a wounded elephant had been left unattended and he voiced his fury. The hunters were not too amused at the slights Colin laid at their feet, but they agreed to

search through the old trophy photos to see if they could find the hunter who had wounded the elephant.

'High brain shot,' the hunters said, 'bullet must have lodged in the soft brain tissue, can't remember which bull it was though.'

Finally the photo was found and Colin's mystery was solved. The man whose shot had wounded the elephant was the spitting image of Colin himself. The elephant was on a payback mission.

The decision had to be made, if Colin and Nancy were to stay then there would be no choice but to shoot the bull. They made the decision to leave, after all it was they who were in the home-range of the bull and they were determined to respect his right to live. (He has since heard that the bull is still living and seems content with life having never moved back to the region of the hunting concession).

Back in the town of Vic Falls, we were persuaded by friends to give a few talks with slides about our trip thus far, and at one of these relaxed bar sessions we met old Mr. Connolly again.

'Well boys,' he said, 'you've made it this far but the hardest section is yet to come. You've got crocs and hippos like you've never dreamed of below here.' He sipped thoughtfully on his cold beer, 'let me know when you're finished. I'll have a beer for you.'

The evening before we left, I fell asleep at the fire and some light fingered tourist stole my hat from my head. That hat had made it this far down

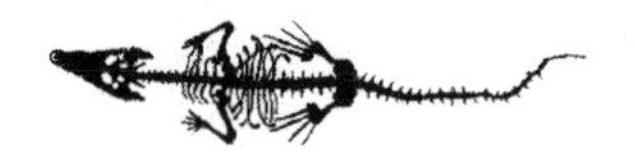

the Zambezi and would make it no further - I was pretty miffed. Still, I was better off than Livingstone who had kilos of valuable trading goods, medicines and necessities stolen by deserting porters all along his route up the Zambezi.

Feeling refreshed, and having built up our weights again with good food, we set forth. In September 2009 the Zambezi was in high flood and the Batoka Gorge and the upper and lower Moemba falls were un-runnable (not that our kayaking skills would have allowed us to run them anyway!) We agreed that seeing as we had run this at the very start of our trip on the Shearwater rafts, we could consider this piece of water complete. Order to us was unimportant.

Mr Reid kindly loaded our kayaks in his car and drove us to the old colonial fishing club at Deka where water access was easy. We unloaded in the shade of a massive apple-thorn acacia and took a full hour to balance the kayaks with our new loads.

We waved another set of goodbyes and pushed off, all of us eager to begin again and to find that solace in the wilderness which we had missed in our week of crowd-mingling up in Vic Falls. Pierre Elliot Trudeau said once, '*what sets a canoeing expedition apart is that it purifies you more rapidly and inescapably than any other travel.*' This rang especially true for the three of us.

Soon we came into thick, harsh thorn-scrub like bush which opened out here and there to reveal a few lonesome lodges. It didn't take long to paddle through this bush to where the water led us into what used to be a gorge, but what was now the backwaters of the Kariba dam. It still retains the name of "Devil's Gorge", and sports high cliffs over

which rock hyraxes scuttled and klipspringers nimbly bounded away from danger. Once you 'got your eye in' they were easy to spot and the rock face literally bopped with them. The water flowed slow and green down the gorge and our going matched the flow as the intense heat sapped our strength and our body fluids. Again and again we gulped down the fresh water from our water bottles until there was none left, and then we simply used the empty canisters to pour the warm river water over our heads.

The heat made visions swim across my memory and I recalled the same intense thirst on a walking expedition that I had joined for some distance along the Suguta Valley in Northern Kenya, when I was about fourteen years old.

The landscape there was slightly different to the snaking river that we paddled on now. The valley is part of the Great Rift Valley that rifts the continent of Africa apart. Its floor is a myriad of sharp lava peaks that tear footwear apart within a day. We had planned a four day camel safari and had packed accordingly. The camels carried two hundred litres of water, twenty litre plastic jerry cans that were strapped to the camels with ropes and lengths of rubber. We woke early, at three a.m., with the intention of walking the full sixty kilometres to the Suguta River by dusk, planning to spend the heat of the day rested up wherever we might be able to find shade.

The valley is harsh and treeless, a lava wasteland where rain only comes once a year spurning a thin blanket of green upon which a few specialised species' eke out a living and camels grow fat.

Sometime into our walk we noticed that the

water was leaking. The lids of five of the jerry cans had worked loose as they rubbed against the loading saddles of the camels. They had been tipped sideways and we had trailed precious water for mile after mile. Our rations severely depleted and our going slow we were forced to camp some thirty kilometres before the river. When the camels had drunk their fill we remained with only some twenty litres.

The next day was a feat of endurance as we strictly rationed the remaining water, our tongues swelling and our faces blistering beneath the unrelenting sun. We sucked on jagged lava as there were no smooth pebbles to be found. This kept our salivary glands active and our mouths from drying up.

The heat was oppressive, its fingers strangling us in a caustic grip. Soon the desire for salt overcame the desire for water and we zigzagged our way over the lava fields and brown dirt looking for anything that might yield salt. Come early afternoon we came upon three locals, young Pokot tribesmen who wore simple *shukkas* of dark purple, tartan blankets wrapped about their obsidian coloured bodies like togas. Their wrists and ankles were bedecked with leather decorations and around their necks hung elaborately shaped metals that were sharpened to kill and sheathed in leather that guarded against injury. They conferred with two of our men who could speak fluent Pokot and soon one man informed us that the tribesmen knew where there was water. It was close by, but they would only show two of us the way.

Josh and I were voted to go and with gesticulations that indicated we should bring only three camels and all the water containers, the Pokot

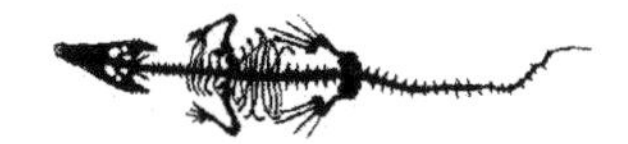

led us across the lava away from the rest of the group. Our pace slowed and we finally stopped beside a large hole down which one of the Pokot began to climb. We followed him in and there lay a deep lava cave that was cool and dark, its walls shadowed and cold to the touch. Some fifty paces into the cave there lay a pool from which we quenched our stifling thirst and splashed our faces and bodies with a coolness that balmed.

I was torn from my reverie by Ace, 'Guys come have a look here,' Ace called from ahead where he appeared to be peering into the bottle green depths. We paddled up to him and looked down into the water. Everywhere there were these freshwater jellyfish. Opaque white against the green of the water they had between thirty and thirty five 'arms' that lazily propelled them nowhere. Of course it was Reilly who discovered that they didn't sting, and we picked them up and examined them closely. I've never seen them anywhere else.

As we paddled deeper into the gorge the blue sky formed a roof high above our heads and *Ficus* trees clung precariously to rocky outcrops. There were green pigeons and falcons, spur fowl and francolin and in one memorable moment we watched a verreaux eagle swoop down out of the blue and pluck a squealing hyrax from the rock wall.

The gorge was a pretty eerie place to be. Each paddle stroke that we took in the green water echoed round the rocks. It made us wonder about the crocodiles that lived here. We never saw one and could only imagine the monsters that were lurking below us.

There were no places to camp in the gorge itself and so we waited till it opened up. As we

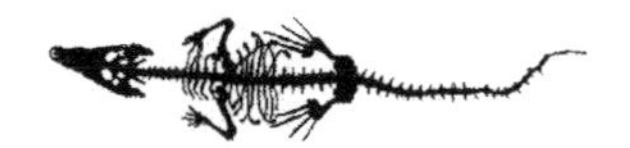

reached the end of the gorge we came across an almost impenetrable floating raft of vegetation that had obviously been blown up the waters of the Kariba Dam to its top end where we were entering: it had now effectively sealed the Zambezi as it entered in. The vegetation was so dense it looked as if we could walk across. It took a good hour to fight our way through, pulling and slashing with our *pangas* for over a kilometre. Finally we reached a clear piece of dry, flat land where, exhausted we could lie in our hammocks and eat a late lunch.

We had met a man in Vic Falls who had told us that on a previous lone safari he had come down this gorge to find a massively bloated, dead elephant floating amongst the flotsam here.

'You wouldn't believe it' he growled through his pipe 'a whole ecosystem had sprung up around the carcass. Vundu (catfish), whole schools of fish, even monitor lizards and by God you should have seen the size of those crocs, ten footers or more each of them, easily. And crazily of all, this big male lion had obviously trailed the smell of the rotting flesh and twice that day he swam across from the mainland to feast.'

But we saw no swimming lion or floating carcasses as we pushed on an hour or so before camping for the night as we all agreed we did not want a devil in camp that night!

Now all of Kariba lay ahead, two hundred and eighty kilometres of hard paddling against high winds and angry waves. We had agreed to meet friends who were spending a week in a houseboat at the far end, and so this was our short-term goal. We knew that in fourteen days we had to be in house-boat territory – we made it in fifteen.

Sketch Map of The Kariba Dam

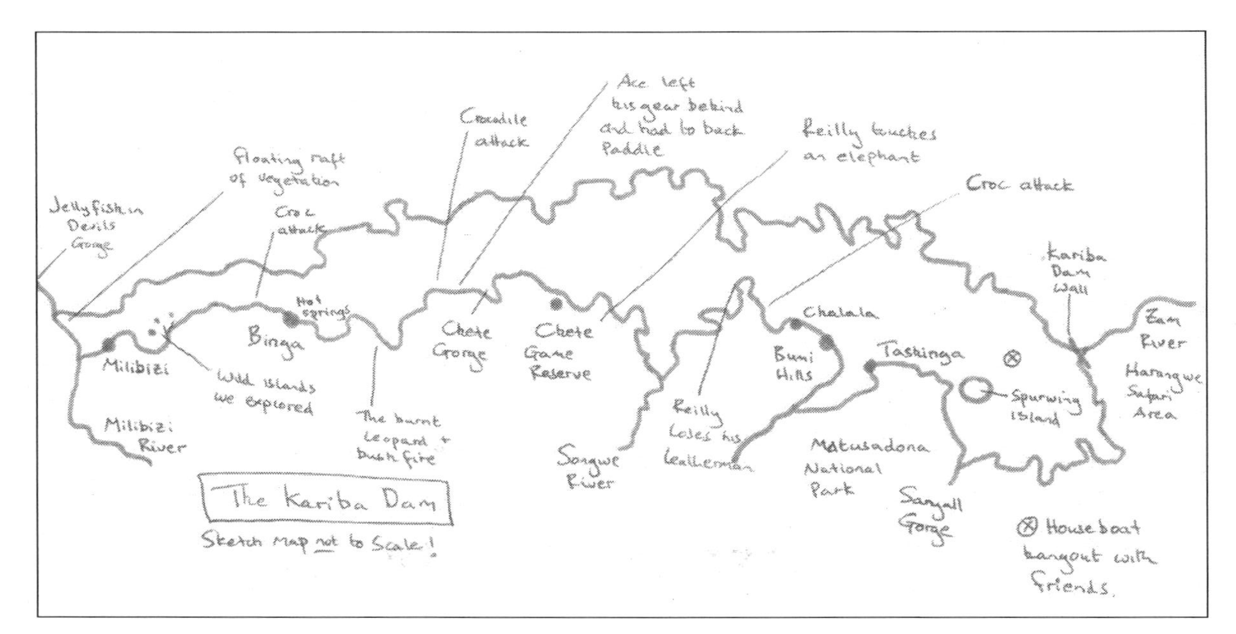

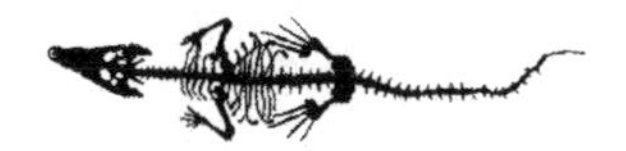

Kariba is the ancestral home to the Tonga tribe who, for nine hundred years, inhabited both banks of the great river from Kariba Gorge upstream to Devil's Gorge, the area now known as the Zambezi Valley. The valley is a hot and harsh land that is described as inhospitable by many, but the rains, from December to March bring fresh, green grazing, that dances in the cool rainy-season breezes on rich alluvium soils. Here wildlife abounded and provided ample hunting. Today wildlife numbers have dropped hugely and many conservation organisations operate around the dam.

In the mid 1950's Tonga history changed. They were forced to resettle high up on the surrounding escarpments, and in 1959 Lake Kariba became one of the largest man-made lakes in the world, with a length of over two hundred and eighty kilometres and a depth of one hundred and sixteen metres. The rock that gave Kariba its name is now more than thirty metres below the water's surface. And this in the eyes of the Tonga people is a cause for great concern.

That rock is the home to "Nyaminyami", the river god, a large half-serpent, half-fish like creature who is the protector of the area. When the plans for the construction of Kariba Dam were announced and the Tonga resettled, they believed that the river god would be so angered that he would make the water boil and the waves rage, so as to destroy the bridges and concrete structures of the insolent white man.

Construction began in 1956 and a year later the mighty Zambezi rose to flood level. Great volumes of water hurled themselves down the narrow Kariba Gorge, and equipment and access roads were destroyed. To the Tonga people, this was "Nyaminyami" in action.

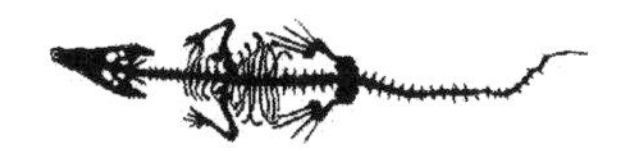

Experts were called in to assess the damage and to make changes to the planned structures. They determined that the floods were a one-in-a-thousand-year occurrence, and made the necessary changes. One year later the mighty river flooded three metres higher than it had done the year before, and again roads and infrastructure were washed clean away. This flood destroyed the access bridge, the coffer dam and parts of the main structure. In this flood three white men were killed and dozens of searches failed to locate the missing bodies for their families, even with the assistance of the Tonga tribesmen who knew the river well.

Weeks of searching found no bodies, and the authorities, in desperation and under pressure from tormented relatives decided to follow the advice of the Tonga elders. A black calf was slaughtered and floated with trepidation on the river.

The next morning the calf was gone and the bodies lay in the immediate area.

The disappearance of the calf is indeed no surprise as many a crocodile could have taken the opportunity to feed, but the *reappearance* of the workers bodies days after they were killed is not so easily explained.

The Tonga believed the dam would never be finished and many had to be forcibly removed from their homelands by the army. The disruption to their life was enormous and the compensation minimal. Today many of them survive as subsistence farmers and supplement their diet with fish from the lake; however poverty and disease remain a constant battle for every family.

The Tonga culture is steeped in witchcraft

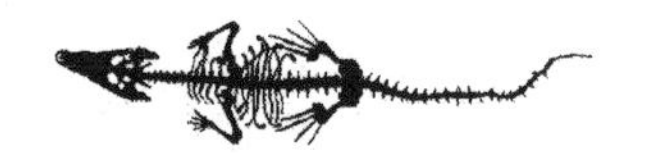

and they resolutely believe that the building of the dam walls separated "Nyaminyami" from his wife, who was visiting relatives downriver when the dam wall went up. "Nyaminyami" will one day get to his wife, they say, and when he does the dam wall will crack open and millions of gallons of water will wreak havoc on the land and peoples below. Often there are small earthquakes that shake the area and these, they say, are the beginnings of "Nyaminyami's" attempts to reach his wife.

In the early stages of Kariba we found islands that were pretty wild, untouched and unvisited. Many of the Kariba islands still boast animals that have either swum across or are descendants of the wildlife that remained when the dam waters rose.

When the dam filled many animals were lost as there had been no plans, by any government or dam contractors put in place to relocate the wildlife. Local farmers came to the rescue, "Operation Noah" was a rescue led by Rupert Fothergill that lasted for four years and relocated over six thousand animals out of harm's way. Elephant, antelope, rhino, leopard, lion, zebra and warthog as well as many small birds and even snakes were taken in boats to the Matusadona National Park.

Tony Bruce, who remembers the 'operation' as if it were yesterday, recalls on a website he has dedicated to tales of the rescue, that all volunteers had to sign a release form that promised not to hold the Queen responsible for any deaths that might occur, and that, should a death occur, the recovery and transport of the corpse must be paid from the accounts of loved ones. This being because Zimbabwe used to be a British Colony.

Topographical maps were used to determine which islands were the most important to clear as the waters rose rapidly. Large, welded metal, open boats were used for transport, often powered by two engines and pulling another smaller vessel. Capture nets were spread out over the islands and any captured animals had their front and back legs bound with nylon stockings from where they were hurried to the waiting boats and loaded until there was no more space.

Boats scurried back and forth to the mainland as quickly as they could to reduce the chances of stress-related deaths amongst the captured wildlife. Once at the shore, the bindings were released and the animals dropped one by one into the water where they swam strongly into the shallows and quickly raced off into the surrounding bush. Sometimes though, confused animals would begin to swim out into the lake and these individuals had to be stewarded back to the shallows by the boats. The water would rise feet every day and often the bloated corpses of animals were found floating, reminders of tragic bids for freedom that ended with exhaustion and drowning.

Rupert Fothergill earned legendary standing amongst those who cared about wildlife. His work as head game ranger of the then National Parks and Wildlife Department in Rhodesia, made him one of the very few white men on the payroll and yet he was accepted as a master of his trade, a man who was respected for his understanding of the wilds of Zimbabwe. Fothergill documented much of his work on an old sixteen millimetre film camera which his grandson is currently in the process of digitising; images that will tell the tale of a daring man who succeeded in a monumental rescue mission.

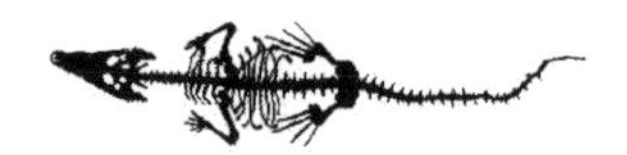

Kariba cost over four hundred and eighty million US dollars and yet barely a penny was spent on wildlife relocation. There is no denying that today the dam remains an important hydroelectric scheme providing power to much of Zambia and Zimbabwe. It is operated by the Zambezi River Authority and is equally owned by both Zambia and Zimbabwe. However will we ever really know the true cost, to the wildlife and to local tribes? Therein lays one of Africa's most precarious balancing acts.

Despite high levels of poaching and recent droughts, all along the river we had seen elephants and buffalo as well as the spoor of lion and leopard, in fact almost the whole 'Big 5' except rhino which sadly had been poached out in times gone by. Only Matusadona remains as a sanctuary for these magnificent creatures.

Early on in Kariba we came across an incident that made our blood boil and made Reilly especially, a raving abuse-thrower – though one without a target as the hunters had already flown home to their luxury Californian lifestyle.

Trophy hunters in search of a leopard had decided to use a phosphorous round, a dangerous munitions round that burns quickly and produces an immediate blanket of smoke. They had hit the leopard but in the process had started a bush fire which, as well as burning the area for miles around, had also charred their trophy. We saw the burnt spots of the leopard and the still smoking bush, as well as the embarrassed concession holder who hurried us away with curses worse than Reilly's.

'Feckin trophy hunters!' Rei shouted, hoping "Nyaminyami" would hear him and inflict revenge somehow.

Each man has his own opinion on hunting and a hunting-conservation debate will fill an entire book and more on its own. Suffice to say that hunting is a sport that fills me with disgust and a revulsion that turns acrid on the tongue even talking about it. I do not have good experiences and perhaps it all stems from my first integration with hunters (Kenya being a non-hunting country). Having just come out of Bush Academy, I managed to score a job in a hunting concession near Caia in Mozambique. As a young lad growing up, many of my elder peers in our community worked as hunters, and even now as we left school many of my friends turned to hunting to earn a living. The money formed a large attraction; wages were good and tips were excellent. I thought that I too would turn to hunting.

Now though, years on, I see trophy hunting as the callus murder of a creature so that the hunter may return home with a 'story to tell' and a mounted head to boast about. It is in my belief that these days, in Africa, there are very few ethical hunters. There are those that claim that they themselves are ethical, but if, in all honesty they cannot prove the ethics of the company they work for, then I do not believe they have a right to justify their own moral claims. If that 'ethical' company works under an un-ethical government then where do they stand morally?

At the end of any shoot, anywhere in the world, genes are removed from the gene pool, and often it is the strongest genes.

I had come across the bull-headedness and sheer ignorance of a hunter in Mozambique. The hunter had arrived as a client holding a permit to shoot a male sable. We searched the concession

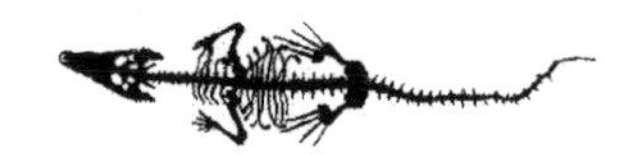

high and low and could not find a sable that I considered viable. One afternoon we came upon our breeding herd whose male was a magnificent specimen of velvet black, with a huge set of horns of over thirty-seven inches that arched back gracefully over his body.

'Wow,' the client breathed, 'can I not take that one?'

'No way!' I was adamant and furious that he should have even enquired. 'That sable is the head of the breeding herd; it is he who will sire next season's young, who in turn will become trophies in their time. The breeding season is in full swing, if you take this male out then you upset nature's balance. The competing males will have to begin to fight for dominance again, and should one win it will be too late for them to sire young as the females will already be out of season and next year's young will be severely reduced in number.'

The man huffed and puffed and muttered rudely under his breath as we drove away. This was his last day on the concession.

I knew nothing more until just after dawn the following day when I saw a large sable male being unloaded from the pickup.

'Where is that from?' I asked the staff.

'*Baas* told us to take that man to get this sable,' they apologised, knowing the stance that I held.

I blew up. The client had purposefully gone behind my back to the concession manager and had demanded he be given a sable trophy. Instead of holding his ground my boss had given in (as he did,

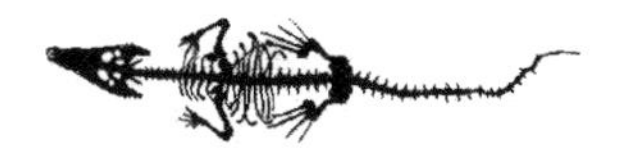

time and time again). In reward he had been tipped handsomely, me not at all, and we were short of a viable breeding male to sire next year's young.

In my fury I quit my job and did not look back.

Back on the river, at the site of the burnt leopard, these thoughts marauded like furious beasts in my mind as in the day, we navigated from island to island trying to avoid the wind and following the southern shore. Often we were forced to paddle at night simply to avoid the rough conditions that came as the winds reached twenty to thirty knots. Our kayaks were battered by the white horses that reared up and slowly they filled with water making the going slow. It was only possible to empty the water when we found a patch of dry island land.

Food was easy to come by. Fruit trees seemed to wait for us at every stop, berries were in plentiful supply, fishing was good and we made biltong (sun dried jerky) from crocodile tails to chew on as we paddled.

'The fish are fat and lazy here with no current to swim against. Every day we are hooking 6 -7 – 8kg tigerfish that fight like soldiers and sometimes, together with the damned wind, take us way off course. We have lost all our lures and have now resorted to making our own. We tie a hook onto the end of our one-and-only spoon, which we then mount to a wooden fish that Ace has carved. When you pull this through the water the tigers go wild.'

Snakes often darted into the water as we

drew up on each shoreline and hyenas called to each other through the dark nights.

Although Ace was organised, he was hopeless at remembering things and on one stop he managed to leave his camera, the GPS and *bhangi* behind.

We had paddled at least ten kilometres and had set up camp when he noticed with dismay that they were missing. M*ascottis* were a vital part of the trip. He would have to go back! At four in the morning, after a few hours sleep, he quietly slipped out of camp with a splash and headed off the way we had paddled in. By mid-morning the next day he was back with us but was exhausted. Ace's fitness and endurance showed how, now, long into the expedition, just how capable we were of achieving the long hours of paddling that were required.

As we reached the middle stages of Kariba, the islands spread further apart and the wind seemed to pick on any one of us and worry us off course. There were a few nights where we did not camp together and had to meet up again in the following days. There was no communication between us on these partings and it was quite by chance that we met again. (When I say chance - it was pretty easy to spot a paddling silhouette on the horizon!)

One afternoon I had stopped to collect wild figs and when I got back on the water found myself alone and far behind the others. I was doomed to what ended up being two nights on my own. But we each carried enough food for exactly this reason, but this time I had both Potty the teapot and the *sufuria* or saucepan, so while I steamed hot, syrupy figs, they ate smoked fish and biltong. Sixty kilometres

later I found them and was given a royal bollocking.

Not many days later, I veered off to paddle into a bay and managed to stir up a female croc that was obviously in the early stages of her nesting.

'Danger! Jamie danger!' Reilly yelled, as the female raced right at me. I could just see the top of her head and I turned and paddled hard out into the lake.

Several times we were mock chased by crocs and they were like torpedoes through the water, so incredibly fast. It was always the croc that ended up stopping though, and each time we paddled like stink till we could no longer see them!

In Binga we called in to visit a crocodile farm to resupply and spent the evening in hot springs that salved our aching muscles. It was Reilly's birthday so we made him a jar of sweet, stewed figs:

We were now hugging the shoreline as we

headed past Chete gorge. Reilly was obsessed with elephants and tracked them whenever he could. Often this would mean either waiting for him for long hours, floating over how many crocs that lurked in the inky waters! Often we went to shore with him as photographers.

One day Reilly awoke with a bombastic announcement:

'Jimbo, Ace, today I'm going to show you how to touch an elephant,' he said.

'How the hell! Reilly, you're such an ass!'

'Yeah you're right I'm going to touch its ass, you just watch me!'

And that was that. Reilly was decided. It was a fine day when we came upon a group of elephants that were swimming. We found that if we sang while we paddled, we were able to get quite close to them, and suspected that Reilly was going to attempt his manoeuvre here. But 'No' he emphatically stated, 'I have to be on foot, on shore. I want to walk right up to him.'

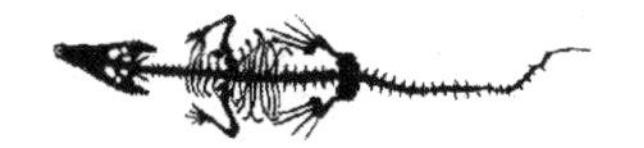

Chapter Eleven

Kariba Dam – Batoka Gorge – Mana Pools

Thou shalt earn thy food and rest rewards after wearing toil

One windy bright day as we set off, a blue sky full of flying clouds, we came within view of a few elephant dotted around the hinterland back from the lake shore. Reilly declared that this was exactly what he wanted. We paddled to the bank and Ace and I watched amused, as first he dived into the water and then rolled about in the sand. All around there were fresh elephant turds and Reilly gave himself a good rub-down with these too!

Camouflage complete, he began, ever so slowly to move forwards. He was relaxed and confident as we had been taught to be at Bush Academy. The elephants paid no heed and continued to browse. They could smell nothing except for perhaps another elephant. As he got closer both Ace and I waited with wide eyes and chewed our lips in nervous anticipation. He worked his way in closer and closer, until he gently put his hand out and patted an elephant on its behind. He had managed it! Reilly was jubilant as he returned with shining eyes; and the elephant? Well he never even noticed!

Ace spent several days fashioning a new fishing rod and lures. His rod was straight and true, the eyelets were from beer caps and epoxy resin. His lures were wrapped in silver paper (aluminium foil and cigarette papers) and dangled a hook behind, a rig on which he proudly caught bream and tigerfish.

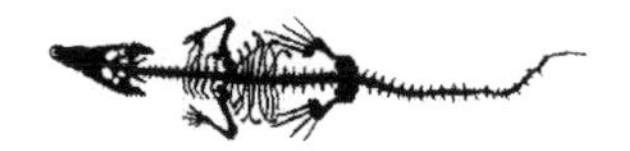

Homemade rods often have a habit of snagging a greater catch than 'professional' rods and Ace's rod reminded me of another Aberdares trip with Dave.

On reaching camp our Aberdare stream-side camp we had discovered that we had forgotten our fishing rods. We spent a merry afternoon lying out on the soft tussock grasses fashioning homemade ones. Using long shafts of bamboo we collected stringy bark with a hint of flexibility that we used to make eyes which we attached with soft wire that we pilfered from the Land Rover engine. Luckily all had not been forgotten and we had line and flies upon which we caught large speckled trout that we honey roasted in ginger leaves upon the glowing coals of an evening fire. It was on this trip too that we caught American Brook trout – a fish that had not been caught since its release in the 1900's by Ewat Grogan.

Back on the Zambezi we drew into Chalala and Bumi Hills at the end of the second week on the lake. We stayed for a few nights in the Matusadona National Park where Imire (Reilly's home) has close links with the Tashinga Initiative, a set-up that works to protect and educate in the local region. Matusadona is where two of Imire's black rhinos have been successfully reintroduced into the wild. We walked for hours tracking and watching our rhino and looking for rock pythons. We were also lucky enough to spend some time with the awesome Bumi Hills Anti-poaching Team who were incredibly inspiring.

One evening as we had sundowners on the lake shore, a huge bull elephant came down to the river and began to gambol in the water; splashing about in it with his feet and shooting glistening jets

of water high into the air and over his back. He was obviously in love with life. And we were too. We smiled with him as the sun sank low over the horizon.

Elephant have a playful side and I remember being very young and watching an event that to this day plays clearly in my mind. In 1993 my godfather was tragically killed by a falling tree. He was laid to rest in the Aberdare Mountains. We attended his memorial a year after his death and one evening a herd of wild elephant passed through camp. One of the guests, Alan Binks, had perfected the deep stomach like rumbling of the African Elephant (*Loxodonta africana*) and he used it now to initiate 'conversation.' I remember one of the wild elephants stopping to listen. She moved closer and flapped her ears. Alan then picked up a smooth, rounded stone and with a stomach rumble that must have meant 'hey catch this!' he rolled it gently at her. The elephant caught it deftly in her trunk and rolled it straight back. For a minute or so man and elephant played a game of catch before belly rumbles indicated that goodbyes were due.

Stories like these were told and re-told and mulled over as, come the next few days we began to see houseboats. Kariba is famous for its houseboats. Most of the houseboats we passed had once tempted thousands of foreign tourists, but these days it is only the resident population that is keeping the industry going. A great many Zambian and Zimbabwean families make time once a year to spend a week on the lake. As we got further into this area, small towns sprang up on the lake shore. We pulled in at a few to get extra supplies, but the economic state in Zimbabwe at the time meant that many of these towns were just ghost towns, trying desperately to survive as the cost of staples

rocketed.

We managed to trade for fish-hooks with staff from the kapenta rigs (more about these later), and, as they fished at night we would ride behind them in their wake and hook monster tigers. I lost two rods in this way. One was simply pulled out of my hand after a fifteen minute fight with a tiger on the end of a line that had a forty kilogram breaking strain. The other simply snapped in half with the fight. We fished in every way possible – with spears and flies, spinners and lures, hand-lines and trawling. Our days were filled with fishing, our evenings with the cooking and eating of our catch.

Char-grilled Tigerfish

Tigerfish fight like demons in trouble.
Real reward for real work.

The flesh of any river fish is often firm and
is good protein for the weary adventurer.
It can be cooked in a multitude of ways
and really is, 'manna from heaven.'

This dish is not just about the preparation but about the catch too. It is best to take a sunny afternoon to fish for tiger, to revel in the fight that he brings to you and then to savour this dish as a main meal at dinner when you are warm round the campfire. Any left-overs can be fried over hot coals in the morning for breakfast.

Serves 3 fishing obsessed, (some might say "possessed"), but *still* ravenous kayakers

1 x 3kg tigerfish
1 x fresh lemon
Pinch of herbs and spices
Pinch of salt

Chilli (oil)

1. Lay your lines and catch at least a three kilo tigerfish.
2. Using your Leatherman, gut but do not scale the fish.
3. Now place your fish on hot coals until the scales are charred.
4. Remove the fish from the coals, and using a flat surface like a rock, carefully remove the scales to find a dense white flesh beneath.
5. Paint a mixture of herbs, spices and salt on the flesh.
6. Add a squeeze of fresh lemon, if you have it, and savour your fine meal.

Often we would pull in on empty, shaded sand banks and would wait out the wind as it shrieked across the water. We had a lot of time for exploring. We would look for crocodile eggs using monitor lizards as our guides, quietly watching them till they found a nest that they would begin to dig up when we then leapt up and scared them away, uncovering their spoils for ourselves.

Often we would split up as each followed their own interests, Reilly stalking elephants and Ace watching birds while I never got bored of fishing. On one of our days exploring we found an enormous python coiled in a tree. It was a magnificent specimen that we this time wisely, left well alone. We had learnt our lesson from the mamba!

That afternoon I regaled the boys with a story of a 'snake encounter' from my time working at a lodge in South Africa.

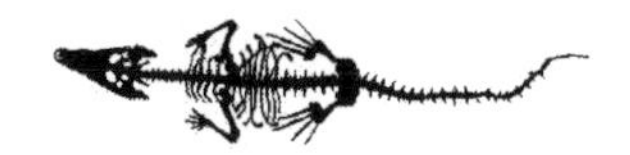

On the lodge's payroll was an old man who was simply too old for work, but as he had been with the staff for twenty years, management found him a job that suited. He was 'the snake man.' Any incident with a snake, any snake sighting, any snake identification we called old Mwenda.

One day a guest noticed the tail of a snake disappearing into the skirting board along the corridor that led out to the open air dining room. Mwenda was called and a small crowd appeared. Through a hole in the skirting board the last bit of the tail still protruded. Mwenda had an old single barrel shotgun which he loaded now as he simultaneously and toothlessly pushed the growing crowd back along the corridor towards the kitchens. Now ready, he took a good hold of the tail with his left hand and heaved. The snake did not budge. Clearly this snake was large and strong. Mwenda tried once again but with no success. He stopped to ponder the situation and as the cogs in his brain turned, a smile lit up his face.

He now laid the shotgun down on the floor and with two hands he heaved backwards. The very large body of a python appeared. Mwenda released one hand to reach for the shotgun and winced as the snake disappeared back into the skirting board, pulling his one arm behind. Again Mwenda tried this manoeuvre and again his sinewy strength was humbled.

Suddenly he had a brain-wave. He smiled at his audience. The crowd gasped as they saw his plan.

He arthritically knelt down and untied a shoelace. He tied the loose ends in a knot to the python's tail. He tested the knot. It would hold.

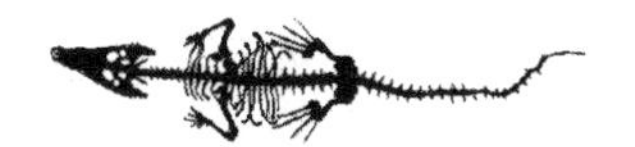

Now he grasped the tail with his left hand again, grunting he pulled with all his might, left hand and left foot arching upwards as, in his right he held the shotgun at the ready.

Suddenly havoc reigned. The snake recoiled and its whole body came loose, it curled backwards on itself and up towards Mwenda's face!

Mwenda visibly paled in fright and dropped the shotgun. He took to his heels and ran. Down the corridor and into the dining room where guests, who were not in on the story, were quietly eating. Mwenda raced between their tables and down the stairs into the lodge gardens. Baboons scattered and kudu leapt six feet high. All the while, as he ran, the python, still tied fast to Mwenda's shoelace, arched gracefully upwards as his foot came out behind, snake-eye to Mwenda's eye, and then slammed down against the ground as his foot touched grass. Each arc threatened python coils over which Mwenda could easily trip but with his nimble footwork he was saved each time.

By the time he had finished running, the snake had been battered to death and Mwenda had learnt his lesson. Tying a snake to your shoelace is not such a fine idea after all!

Back to the Zambezi and Reilly discovered he had lost his Leatherman. There was no way he was leaving camp without it. He claimed he had last seen it on the sand just minutes before. And so we began sifting through the sand. One hour – two hours – three hours later and we had sifted a massive heap of sand through the hammock. We broke for tea and then turned back to sifting. About four in the afternoon Reilly sheepishly piped up,

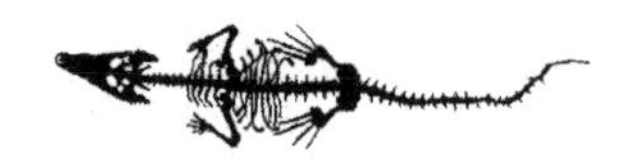

'Got it boys.'

'Where was it?' we asked

'In the spice bag,' he shamefacedly replied.

His reply earned him a good few fresh elephant turds on the face!

One night we strung our hammocks on the rather muddied lawn of an old abandoned fishing lodge. About ten that night I woke to shuffling. I opened one eye and saw a black shape moving towards our hammocks. I shone the torch gently through my fingers and there was this minute baby hippo. I worried that a startled hippo might miscalculate a run, ploughing through a hammock or two. I woke Ace and Reilly to warn them not to get up to take a leak. My voice startled the hippo slightly so I waited.

'Reilly, Ace it's pretty dangerous having this mother and its baby so close, see how small that baby is?'

They both agreed.

'So what's your plan?' whispered Ace.

I had no plan and we sat in silence for minutes watching this female graze closer and closer.

'OK on the count of three, leap out your hammock and into the tree and hopefully she will head in the other direction. One – two – three –'

We all leapt into the trees above and, as planned, the mother and her baby disappeared towards the water and we didn't see them again that night.

We often had hippo in our camp at night as we were so close to the water's edge. Having them come too close was a bit frightening so periodically we would scare them further away which meant waking in the night.

After our night-fishing escapades we could come out of the water unannounced, blocking the escape route of a startled hippo. As a result we always had to be exceptionally careful, flashing our torches up and down the banks and talking loudly so as to alert any hippo of our presence.

Remember that hippo that treed us in the dam in Zim with a bottle of whiskey? I told you about him earlier in the book. We'll go back to him now. He had become an 'agro' hippo determined to inflict damage and the decision was taken to remove him. In one night foray he had flattened a large section of maize upon a farm that a particularly nasty war vet had taken over. This wicked character rounded up his cronies and together they marched onto Imire demanding monetary compensation. We carefully dissuaded them and said instead that we would kill the hippo (we already knew it had been wounded that fateful night) and would donate the meat to his cause. The proposition calmed him.

'Jamie, Reilly - job's yours' John Travers said and so once again we loaded our trusty little rowing boat, Reilly's .458 shotgun, my .375, and our paddles.

We knew that the hippo spent his idle daylight hours at the northern end of the dam and it was there that we headed. The plan of action being this: Any wild animal has three distances to which it behaves differently.

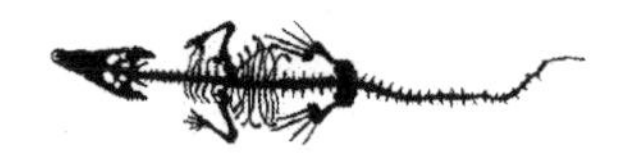

At about sixty metres from this hippo we would enter its comfort zone where it would pay no attention to our actions. It was within this zone that we would prepare for action, loading the weapons and taking position.

Paddling closer we would enter its alert zone at about fifty metres. Here we expected the hippo to watch us carefully for signs of what the next action may be.

Paddling even closer we would enter the danger zone where any normal animal would begin to display threatening signs and would change its whole body language to that of aggression. It is in this zone that the unpredictability of a wounded animal may have lethal consequences. If our hippo should attack here, the plan was the same as that, had we have entered the critical zone (closer still).

At ten metres we would consecutively fire, Reilly first and me soon after. Only a brain shot would down a charging hippo.

There was no option to miss if we valued our lives (and we did!).

Everything went to plan and the hippo charged, its snuff-like bellow curdling our blood through which adrenalin charged. Forty metres - Twenty Metres -

At ten metres, two shots echoed across the dam and the charging hippo fell, its death splash causing a mini-tidal wave which set our boat a-rocking.

Together it took twenty men to drag the hippo from the shallows to the shore where we cut it open. Lodged in its body we found seven bullets that

were festering in open wounds around which we carefully cut. We then distributed the meat – a deep red meat with surprisingly little fat. Most of hippo fat is found in a layer between the three centimetre thick skin and its muscle.

It was this story that played out in my mind as one afternoon I found myself paddling a little behind Reilly and Ace; they were about three hundred metres in front of me. As I swept my eyes across the water, I saw a croc begin to submerge itself as it intently focused on the boys. Sure enough it began to follow them. I was too far back for my shouts to be understood so I had to paddle with all my might to get close enough to them.

'Croc! Croc guys! Danger, croc!' By now the croc was about fifteen metres from them and closing in, and that's when the boys understood my shouts. They turned to see the croc coming at them at full speed. They paddled as fast as they could manage to the bank and leapt out of their kayaks. I caught up and clambered onto the bank up-river from them.

We were safe! Whew. We gave the croc a good hour and a half before skittishly slipping back into our kayaks and paddling out of the area.

It had been thirteen days and we knew that sooner or later the dam wall would come into sight and that somewhere, our *rafikis* were on the water looking for us. Ace suddenly got a rush of energy; perhaps he was thinking of all the good food that lay waiting, and paddled off quite far ahead. By late afternoon Reilly and I were thinking of making camp and had given up on seeing Ace again that night, when the noise of a speedboat filled the air. It was far off but soon we could see it reflecting the last rays of the sun. It came closer and closer and then

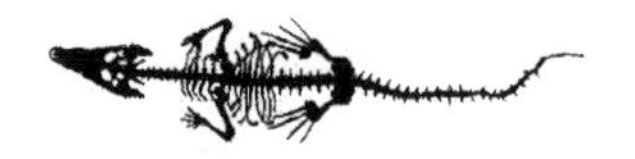

circled us.

It was Ace and our *rafikis*.

We tied our kayaks with relief, to the speedboat and they towed us the four or five kilometres to the houseboat where we settled in quickly with cold beers and tasty nibbles.

In all, we spent four nights partying hard, sleeping off heavy hangovers each day, swimming, story- telling and eating. Unknown at the time to Reilly, it was here, six years later at a similar party that he would propose to Candice.

When the time came to leave we dragged our feet but Ace got us re-motivated and we paddled to an obligatory stop as close to the dam wall as we could. From there we used the car of a *rafiki* to get us to the gorge below Kariba. The roads were horrific and to get the car down to the water's edge was tough going. There were places where we had to remake the road.

Once we unloaded, we had to help push the car back up the steep hill before we could set off in our kayaks to find our first camp in the gorge below the flood gates.

It is quite inspiring to look up at the dam wall from below and realise quite how huge an engineering feat it is, and to realise just how enormous the barrier between "Nyaminyami" and his wife is. The Kariba dam controls forty per cent of the total run-off of the Zambezi, thus changing the down-stream ecology dramatically. The Kariba Dam wall was designed by a French engineer and inventor; a Mr. Andre Coyne. He was a specialist in 'arch dams' personally designing over fifty-five of these in his life. As we gazed up at Kariba we could

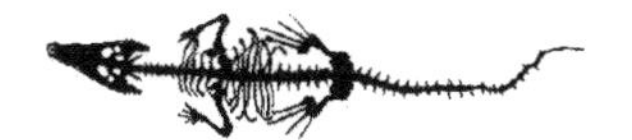

see the genius of his mind. Rumour in the area tells that Andre committed suicide after some of the dams he built collapsed ... a sign the people say, that Kariba too will one day collapse. I have never been able to find the truth behind this much uttered story!

The gorge below Kariba was of dry, rocky beauty and all along, dotted here and there, were plaques commemorating people who had died either on the water or had been taken by crocs or hippos. There were those too who had died in elephant and buffalo attacks and others whose plaques did not state how they met their end. We were not too sure why this gorge was chosen as the site of these memorials, but each was a strong reminder that life is short and precious. There was evidence of heavy fish poaching in the gorge and wildlife was sparse.

Scenically though, the gorge was awesomely rich. Verreaux eagles swung low across the cliffs, and we eagerly searched for the taita falcon although we found no sign of it. There were one or two fishing villages but essentially we were on our own. We camped on one of the islands in the river as the escarpments receded and a cool breeze whipped up.

As we paddled down the river, one side Zimbabwe and the other Zambia, we pulled in at a Zambian fishing village and were saddened to meet wholehearted disinterest (for want of better words) in the rhino. It really was incredibly depressing to come across this amongst the very people to whom the rhino was heritage.

As the escarpments fell away, the wildlife from the Harangwa Safari Area had better river access. Wildlife became more and more apparent on

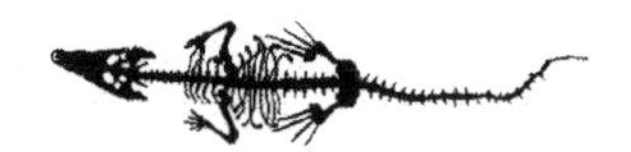

the Zimbabwe banks after the Kafue river confluence and Kanyemba Bush Camp but poaching from the Zambian side was epic. Everywhere we found carcasses and in several instances, the drag-marks of wounded elephant. Locals told us that easily forty hippos a night were being shot. Their meat put on sale in local markets and their teeth sold as ivory.

There were bullet shells all along the riverbank and often we tracked poachers, finding their hides, their meat smoking racks and litter. Once or twice we had to dodge poachers who were well armed and, we even found a dried pair of baby elephant feet on the Zimbabwean side, that had been used by the poachers to tie to their shoes and so cover their tracks.

In this region, especially in Mana Pools National Park (that lay ahead of us), a shoot-on-site policy stands for all poachers. There, they believe, this is now the only effective means left in preserving our wildlife.

This region of the Zambezi though, is prolific with tsetse flies and this at least stops the hoards of livestock herders who tend to drive their beasts into grassed areas, protected or not. Tsetses are a biting fly, rather similar to a horse fly. All species feed on the blood of vertebrate animals, but some are the vectors for trypanosomiases or sleeping sickness. These flies take a heavy toll on livestock and traditionally their presence prevented the establishment of permanent farming and population centres. As a result wildlife was able to flourish. Where the tsetse has been eradicated, Man has taken a firm hold.

Tsetses breed in trees and bush and many colonial policies were aimed at eradicating this fly.

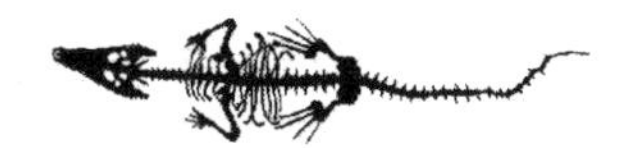

To do this, vast tracts of bush were cut down and burnt and thousands of hectares of habitat were lost. Wildlife also was slaughtered throughout Africa, as it was believed that by removing the animals the tsetse feeds on, the fly numbers would decline. Now pesticides and trapping are the most common control methods used.

Wherever the tsetse is ensconced, wildlife will thrive!

'Ow! This area is definitely a wildlife oasis' Ace exclaimed as he slapped at another of the biting flies.

We pulled in one day to a particularly shaded looking grove of trees behind a sandbar where yellow billed storks squawked and squealed, clattering their bills at us, and happened upon Gerry von Memerty's plaque, a particularly famous and iconic Zimbabwean bush-man who had been gored and killed by elephant there. He had been a good friend of Norman Travers, Reilly's grandfather, and he loved the tsetse. We took a moment to remember the Africa we had missed out on, the Africa that teemed with game and was mostly unexplored.

There remain many hunting blocks in the area although there has been a reduction in hunting clients due to the economic situation. This drop in clients has meant that less money is available for conservation, and now more and more poachers are accessing these once controlled areas.

In Africa, hunting and conservation are inextricably linked. Conservation has become increasingly important as more and more species are added to the endangered, or even extinct, lists.

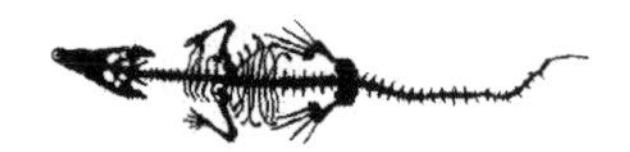

Trophy hunting is a huge market that attracts people from all over the globe. It provides a source of income and employment in remote areas where it is desperately needed, it gives local communities a reason to protect their wildlife and it conserves habitat. And if habitat is conserved species can always be reintroduced, gene pools built up and ecological grandeur restored.

Many a time you will hear it said that the hunter is the best conservationist that there is. This is true if the hunter is ethical and moral – meaning he sustainably manages his hunting block with the survival of species in mind. After all, hunting is his business and therefore conservation is in his best interest. If, on the other hand he mismanages his block, does not follow the strict quota system that is linked to the land's carrying capacity and to the number of animals there already, as well as the balance of species, then conservation is not happening. These operations are often fly-by-night set-ups, there only to make a quick buck like in any other business sector.

A well-run hunting block may well be the answer to conservation in Africa as wildlife pays for its keep (consumptive utilisation) and the hunting operation creates a presence in a region that would otherwise be decimated by poachers.

Many hunting blocks lie far from the trodden tourist route and would not otherwise be visited or protected if it were not for hunting. All African governments struggle to find a working budget for their wildlife departments. If a hunting outfit will cover the patrol costs for a vast tract of land, then it leaves Government departments able to concentrate their forces, more effectively, in a smaller area. In this way, in a perfect world with no corruption,

hunting and conservation can work well together.

That is the argument that one will often hear but experience has led me not to believe this. I truly believe that until trophy hunting is banned throughout Africa we will remain far from the goal of ultimate wildlife protection. How can we preach wildlife conservation amongst the native Africans, of whom many have never even seen the animals we, as higher earners, are lucky to see often, when we set up businesses that are based on shipping in tourists so as they can 'ethically' kill trophy species? Are we really sending out the right message here?

Most hunting companies will tell you that they channel money into local communities to build schools and clinics and that, by their operation being in the region, they bring employment that would not otherwise be available. I do not deny that this does happen with respectable hunting outfits, but so too would it happen with photographic and walking safaris, game tracking courses, veterinary darting safaris and simple vehicle based tourism. There is absolutely no need for animals to be hunted in the name of conservation, but where we stand now we do understand the important presence of controlled hunting in a region, such as many of those that we journeyed through.

Back on the river tsetses made full use of every open patch of skin until we had welts all over our now tanned bodies. The region was so rough and beautiful in a spiny, thorned way that we stayed at one camp for several days and nights. We thoroughly explored the surrounding region walking up to ten kilometres inland, dodging grazing herds of buffalo, watching entranced as elephants balanced carefully on their hind legs to reach into the *albidda* trees to snatch at the curled, browning pods. Zebra

barked warnings and disappeared in hoofs and stripes and giraffe look maturely down their long noses at us. (Or was it their necks?)

The Zambezi continued its watery call and as we left our camp and its dappled shade behind we came upon huge, slate-grey baobabs, hundreds of hippos, artfully wind-blown sand dunes and sand banks, and wide, open sections of river that had low tree-lined banks. Erosion was evident where the animals collected to drink at the water's edge. A lizard buzzard snatched a sunning snake and we watched it carry it up to a treetop where it devoured it fully.

As we neared Chirundu, we came upon a section of bank that had collapsed and been washed away in high flood, perfectly severing a tall termite mound in half. It was incredible to have the mound open to us like a book, to see the passages and the termites going about their repair business. A termite mound structure is an extensive system of tunnels and conduits that form a highly effective ventilation system for the subterranean nest. The nest is a sphereoidal structure composed of cool galleries and chambers where a strict social caste of ants lives out their lives. Below the nest lies the cellar to which the air shafts lead. Termite mounds are conduits for nitrogen from the air to nitrates in the ground and are important in any balanced ecosystem.

Suddenly surprise was upon us. My finger lay on the camera shutter, my eyes watching for the next move of a small falcon. Ace and Reilly were shouting distance in front and to the side of me. We were both drawing to the apex of a small island that we had paddled on either side of when my eye caught the movement of a large hippo in the island reed banks that I had managed to paddle upwind of.

Ace and Reilly must not be able to see it!

I gasped and dropped the camera. The instant it caught my scent, it galloped down from the reed bed, opened its yawning pink mouth and came straight for me. It was too close for me to have any other reaction than self defence! As it reached me, I lifted my paddle out in front of me and, as it happened, hippo and paddle collided which gave me power for a big push off.

My kayak shot backwards and, as the hippo submerged itself a large 'bow' wave caught up with my kayak and I lost my balance and flipped. As the kayak rolled over I saw that both Reilly and Ace had seen my distress and were paddling fast to get to me. My mind told me to be calm, so I hung on, upside down with a frantically beating heart hoping I'd get a T-rescue.... but neither Rei nor Ace nudged my boat.

My air was running out, I would have to pull deck and swim. But where was the hippo?

Some days before we had paddled past a warring tribe of vervet monkeys and had laughed when one was knocked clean off his perch into the water. We were mid river avoiding the crocodiles that lay, eyes and snouts above the water, near where the vervet had fallen. The vervet reacted surprisingly by diving deep and swimming down with the current. Some ways down the vervet came up for air and then appeared to dive deep once again. In this way the vervet reached the shore and scampered out.

In my fearful mind I knew that vervets must know what should be done in such a situation and so I copied the vervet.

I pulled my deck, came up for a breath and dived deep propelling myself strongly as I let the current drag me downstream. In this way I managed to reach shore and, just like the vervet, scampered far from the water's edge immediately.

As I sat shore bound, shivering in fright and watching Rei and Ace rescue my kayak and all the parts that floated about it, the hippo was no-where to be seen. It seemed the hippo had never followed through with her charge. Perhaps the whole terrifying charade was a fright reaction to my having stumbled across her.

Ace and Reilly had managed to flip my kayak upright and were paddling in towards me with my boat in tow. I was awash with the tingles and the bonhomie of still being alive, but even so, I had given myself a massive fright and my heart was still hammering hard as we unpacked and began the long process of drying kit.

After a break of a few hours, a light lunch and mostly dry kit, we pushed onto the water again. Coming up, just a kilometre or so away, was the road-bridge at Chirundu where we had stopped on the drive up to the source. None of us intended to stop at this border post. Plans were to paddle calmly under the bridge just as we had paddled calmly past Kazungula.

Although our ruse worked it was very nearly disrupted by Reilly, who, having spotted baboons sitting on the lower steel girders of the bridge, became determined to knock one off with his catapult. Baboons have tough hides and would barely feel a stone hitting them. It would be the fright that would send them toppling into the river, but, as strong swimmers, they would be on the bank

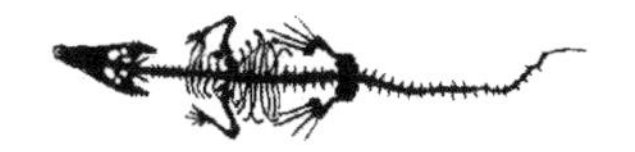

in seconds.

I was in no mood for more shenanigans and Ace was with me on that. We hurried on by and left Reilly to his highly unsuccessful game until we were well out of sight of the bridge.

It didn't take long for the wilds to close in after the busy border post and the rumbling trucks. We were soon to pass the Urungwe Safari Area before heading into the designated Mana Pools National Park, one of Zimbabwe's finest and most iconic parks. It was here in Mana Pools National Park that, 'somewhere along the river' we were to meet once again with our families.

The Zambezi escarpment rose high and blue to our south as we pushed on deeper and deeper into the bush; *Acacia albiddas* hung heavy with pods in *Brachistegia* and *Mopane* woodlands. We heard night choruses like never before; sawing leopards followed barking jackals, followed roaring lions and, here and there, the grunting of hippos slotted in with the whooping of hyenas. We always camped on the Zimbabwean side of the river as it seemed more beautiful, although probably that view was only in solidarity for our country. The river opened up and slowed down, channels meandered around islands of waving riffa palms and there were huge sections of massive forest. This was the Zambezi Valley; the wild west of Africa.

One day Reilly and I were ahead. That morning we had decided that there was to be a prize for the first leopard spotted, and so we were avidly searching the riparian bush when the two of us spotted three male lions feeding on a kill. They were on the top of a mud bank about five yards from the edge. Quickly Reilly and I paddled into the lee of the

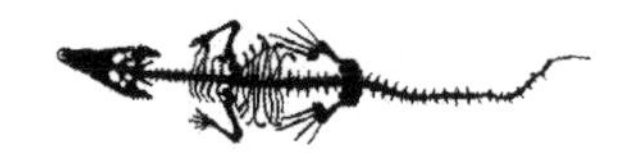

bank and beckoned the unsuspecting Ace closer, signalling him to be silent.

'Ace, come we want to show you something. Come look.'

Ace's eyes brightened and he quickly paddled to our sides.

'There's a massive eagle on a carcass on the bank. We saw it from out on the river. I don't think it's seen us' Reilly whispered.

'Climb up the bank with your camera and get a picture.'

Ace was the only one with a camera now and he fell for the ploy without even a question. He climbed out of his boat and climbed up the bank as we had suggested. As he popped his head over the bank he came face to face with these three surprised lions. We waited. He uttered no sound, he had frozen and a white pallor had come over his face. His eyes were wide and he was totally mesmerised. Reilly couldn't resist and he too climbed the bank and popped his head over the edge. The lions began to growl and soon, in response to Reilly, who was now waving his arms and whooping, they abandoned their kill and ran growling and grunting into the undergrowth some distance away, their bellies swinging low. Ace managed to get his knees back in order and very kindly asked us,

'Please don't do that again.'

We made tea stops under outstretched Jackalberry trees or fig trees where often we would encounter mock charges from hippo as they moved away from our approach. Despite it being daytime, many hippos were out grazing. The males had

begun territorial disputes which made our kayaking more of an occupational hazard than ever before. We tried very hard now to stay together. Choosing campsites was especially difficult as everywhere there were game tracks. We could not hang our hammocks on any tree that might block the routes the wildlife used to get down to the water, or near anywhere where there had been a recent kill, or near a territorial marking post and these could be anything from a clawing post of a leopard to the midden pile of a number of different species. Hammocks do not offer the safest protection to an inquisitive night visitor. One night we discovered that we had made a very big mistake indeed.

We had found a shady, green clearing with a river view and access for a cool breeze. There was nothing around that suggested it would not be the most perfect camp that we had enjoyed yet.

'*Baggsi* these two trees.' Ace stood arms outstretched indicating two especially perfect hammock anchors.

'I'll go off the back of your tree to this over here,' Rei was in there too.

Each of us tied our hammocks and settled in for a good session around the campfire. A baboon barked tiredly somewhere high up in his safe tree as finally, about midnight we turned in. Our hammocks swayed gently as through the night a soft breeze rained *albidda* pods down over our mosquito nets. The pod plops were interspersed with the noise of the water as it rippled past river debris, and now and again, the great big sploshing noise of a hippo ambling upstream against the current in the shallows would flow through the camp.

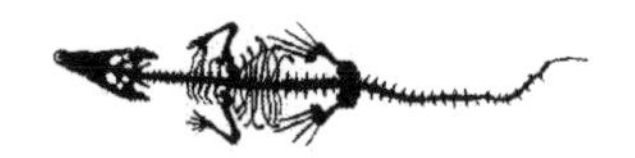

I hear Rei waking Ace in the early pre-dawn light.

'Ace,' Rei was whispering 'look at all the Ellies!'

I roused myself and saw the boys shift their weight in their hammocks to get a good view. Outside the white nylon of our mosquito net ambled massive shapes; shapes with trunks and lazily flapping ears. I breathed out slowly as one female reached up only metres from me and my hammock and curled her trunk about a tasty bunch of *albidda* pods.

But it was not the *hanging* pods they were after. The night breeze had knocked a feast of pods to the ground and as we lay still swinging in our hammocks, quietly having no choice but to be ensconced in the moment, the elephant shuffled their dexterous trunks about the ground scuffling up pods, pods and more pods. Their trunks moved carefully beneath and around us for some time until they moved a few trees over.

We decided the safest thing was to creep quietly away and to come back when the ellies had gorged themselves and left. The ellies were so focused on their meal that not a single individual acknowledged our departure. We sat fifty metres away, squatting on our haunches, waiting, so that we too could have breakfast.

Wildlife has a habit of invading mealtimes but often their interferences can prove quite comical. Many years ago the folks had rented a small cottage at the ocean at Diani in Kenya. Overnight I had prepped potent butterfly bait for my nets; overripe bananas mushed in with generous slugs of whiskey,

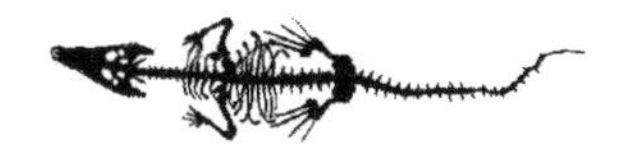

cane spirit and any other booze I had been able to lay my twelve year old hands on. The bait was stored in glass jam jars and already three had burst in the night sending shards of glass and foul smelling mush across the room and up the walls. Mum was none too happy.

'For goodness sake Jamie, put that bait outside in your nets or *something*, but get it out of here.'

I obliged by moving the jars outside on the lawn but before I could begin filling my butterfly traps breakfast was called. As we ate breakfast, high up on a verandah that was built to catch sea breezes, the colobus began to appear. Monkeys that lack a true thumb, their coats hang shaggy and long in black and white that make them seem almost like little old men.

A family of colobus now collected around my smelly jam jars, their nostrils twitched and they chirped at one another in wonder. A large male overcame his fear and reached out to take a small jar onto his lap.

Like monkeys too, we watched as he struggled with the screw-top but the pressure inside had loosened the lid and soon it popped off with a ping that made him run. Soon though, they were all back, the smell was ripe and their monkey taste-buds were salivating. Over twenty minutes they managed to eat two jars worth of bait and therein lay the problem; a problem that we were anticipating with excitement.

It didn't take long for the alcohol to kick in and soon we had a family of rather drunk colobus on the lawn that lurched from side to side. Baby could

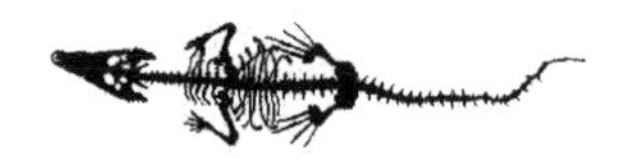

not hold onto Mum's belly and kept falling in a heap between her stumbling feet, little monkey hands had great trouble holding firmly to tree branches, while little monkey eyes looked at us cockeyed beneath their mane of shaggy hair. After half an hour of hilarious cavorting most fell into deep sleeps holding branches firmly with both hands.

But there were no Colobus in this part of Africa, only vervet monkeys and baboons glared, beady eyed at us as we paddled past. We were lucky enough to be invited to stay in a small thatched lodge on the banks of the river by a young couple who we spied breakfasting on the high bank and paddled over to.

Here in their camp, elephant gently picked their way right between the thatched cottages, raising their trunks to determine who was in camp. They had no fear and meant no harm as they plodded amongst the acacias reaching high for the pods.

The view from the lodge looked out across the river valley, cropped green grasses with clumps of golden feather topped sedges. Great herds of buffalo could be seen, black sprinkles on the land, and behind them arose the cloud-frosted hills of the eastern escarpment. Here too we were taken on game-drives by the manager's of the camp and were lucky to see wild dog and porcupine amongst others.

But all good things must be left at some point, and after a few nights we decided that family was calling and so we said our farewells and paddled on. A day or two later we struggled to find a flat, safe and shady campsite and eventually had to settle on a small, cramped space where hammock trees had to be shared. As Rei lit the fire and I shuffled

about baiting night lines, Ace stood and looked about him. As he spoke there was a cacophony of growls.

'Shit those are pretty close,' Ace muttered.

Soon a lion called. Very close indeed. We were all beginning to get the *heebijeebies* now. Again and again lions called to each other.

'What are we going to do?' Reilly asked.

'Let's get into the trees, it's the safest place' Ace suggested.

Quickly we located ourselves a tree and shinnied up, thorns scratching our inner legs and biting into our hands. Ace's sapling leaned over slightly with his weight.

As I glanced at the others another growl echoed and Reilly replied with something that only Reilly would ever reply with; the noise of a squealing pig.

Before he had even finished his call the lions were bounding into camp while Reilly chortled to himself.

It is at this point that I should perhaps explain something about Reilly. Reilly had two party tricks, one that he had perfected after many hours alone in his room as a child, and one which was a risky manoeuvre that might, eventually one day, backfire ... although here's hoping that it never does. His squealing pig noise was so exact that it never failed to draw in the lion or hyenas on any game-drive. His second caper was all brawn. If we ever chanced on lions on a game-drive and the situation was right, before you could even begin to

dissuade him, Reilly would jump from the vehicle and give chase to the nearest male lion. Both of these party tricks were popular and both became his signature moves.

The lions that came now were exceptionally confused. They knew they had heard a pig but now they were confronted with three "pigs" stuck up three trees and emanating a very strange scent indeed.

'Reilly you twat!' Ace hissed, absolutely terrified and not ready to handle another lion incident so soon after the last.

'Uh yees, maybe that wasn't a good idea' Rei relented.

The leaves shook in the trees as a small breeze came through. In my torchlight one lioness cocked her head and looked up at Reilly.

'Now what?' Ace hissed again.

'Sit it out,' I murmured into the dark. I too was silently nervous.

We sat it out for half an hour before we deemed it safe to at least get into our hammocks. Somehow Reilly and I fell asleep while Ace, he told us in the morning, had hung rigid in his hammock until the lion moved far enough away for him to stoke up the fire, where he sat all night staring out the lions who were utterly fascinated by us and probably by the smell of his *mascotti*.

In the morning when we awoke, there was Ace fast asleep sitting upright by the fire and, metres away were five lion, sprawled out and relaxed in the bush. It certainly was a story to tell

our grandchildren that one, although I have to say that Rei has now grown out of his lion chasing tricks!

Years ago we had a different sort of run-in with lions. I was fourteen and we were all off camping - the folks, my sister Hollie and her two *rafikis* who were out from the UK for the first time. We headed up north to the Northern Frontier District of Kenya (NFD) and its vast sand rivers that waved with doum palms while elephants wandered two hundred to a herd.

At that time there was a certain level of instability in the NFD and we were required to hire an armed guard. Ours was a cheerful man whom we named Pistachio in jest.

Pistachio smiled and told us firmly that, 'people like you need to explore places that tourists don't go. I know a place,' he smiled a smile of conspiracy, 'where there are deep clear pools in which you can sit while the fish nibble you gently.'

We loved his stories and agreed that we would head there in the morning. The next day we lazed in the very same crystal clear pools beside the Uaso Nyiro River filled with honking hippos and teeth-gnashing crocodiles. Come midday Dad decided that we should move the car into the shade. Hollie nipped into the driver's seat and turned the key. Nothing happened.

'Oh well,' Dad called from his comfortable recliner rock in the pool, 'no worries.'

After a late afternoon tea where we were joined by a herd of fifteen ambling elephants, we climbed into the car, rolled closed the sliding door and waited. Nothing happened.

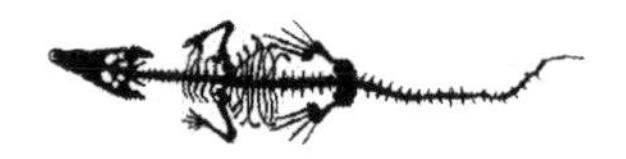

'Battery is dead' Dad exclaimed. 'Jamie hop in the back will you *sonno* and get the spare.' I rummaged about in the back.

'Dad there is no battery.'

We were out of luck. The spare battery had been left behind ten kilometres away at our camp in Shaba National Park.

The folks debated and finally they came to a reluctant agreement. Pistachio was instructed to take good care of me as we set off together to walk the ten kilometres back to camp. Much to Mum's chagrin we were walking through bush where we would be pitted against elephant, lion and buffalo.

Pistachio shouldered his AK47 and I stepped in behind him. It took us several hours to reach camp but we reached there with no great stories to tell.

Quickly Pistachio cut a long pole from a small thorn sapling and deftly removed all spikes with his sharp *panga*. We shouldered the battery between us on this pole and set off for the walk back to the pools. By now darkness was falling and the night sounds were beginning. I was more than a little anxious to reach the safety of the car as thorns reached out and grabbed at us, willing us to stay a little longer in the darkness hours.

Finally we came upon camp where the others had lit a small fire around which they were crouched. I motioned Pistachio to silence and imitated a snorting buffalo as I charged into camp!

Pandemonium reigned!

Five people ran for the two open doors of the

car where none could get in too easily, before they realised it was me.

'Finally,' they laughed. 'The mosquitoes are so bad you can't sit still for longer than a few seconds.'

We fitted the spare battery, climbed in and ... well nothing.

The spare battery was dead too.

There was nothing for it now but to stay put until morning. The mosquitoes whined happily about us, feasting on fresh blood for there was too much flesh uncovered to repel them. We each made ourselves as comfortable as possible and piled ellie poo onto the fire in the hope that the smoke would repel the feasting mosquitoes but to no avail. Eventually we lay in the fish pools where only our faces could be bitten – and they were.

The worry was that we would be seen by a group of *shifta* or armed-poacher bandits, who, (and we had heard many a story), would take pains to rob us of all we had.

Come three in the morning I heard a far off sound and I ran some sixty metres into the bush where I released a flare that burned red as it shot high into the night sky. Our hearts beat nervously as we wondered who it might attract. Soon a small, bulky tipper-truck pulled in. Upon it hung four men – in suits! A further two suited men leapt from the cab and berated us with smiles.

'We have been looking for you *alllll* night.' The smiling man dragged out his 'L's' as he shook each of our hands as if we were long lost friends. We still were unclear as to who these men were.

'We knew that you were gone, and we have been looking everywhere, just everywhere!' another man said, 'what is the problem?'

'Battery,' both Dad and I said together.

'Ahhh, no problem' he replied as he busily began to unpack some jump cables from the cab. We jump started the car and followed the truck haltingly back along the road. At the park gate we said our thank-yous and headed our separate ways, the truck to park HQ and us to our camp where we dreamed of hot mugs of tea, warm beds and mosquito free tents.

It was not to be so.

We arrived to find our camp already taken. A pride of seven lions lay amongst the tents and the picnic table. We tried every which way to move them; we revved the car engine, we chased them with the car, we shouted and we sang. Nothing made any difference, they did not budge. It was not until dawn that the lions decided to sashay from the camp, looking back over tawny pelts with disdain as we fell with exhaustion into our beds.

Our Zambezi lion tale was told again and again as we paddled on down the river, closer and closer to good food and family. Often we came across elephants, big softly lumbering skins of grey that could no longer wait for the wind to knock the *albidda* pods from the trees. They would prop their two front feet as far up a tree's trunk as they could reach, and would stretch their own trunks as far as they could go in the hope of pulling down a tasty snack. When we saw them we would stop paddling and watch. This elephant behaviour never ceased to enthral and amuse us.

Sometimes the river braided into shallow muddy channels that were essentially just wide hippo canals through the undergrowth. Here we often found freckled hippos with creased faces popping up right next to our boats. When this happened we would hurriedly paddle onwards as the hippo dipped under the water. As we drifted past yet another herd of buffalo Ace's voice carried back to us.

'Yeehaa guys, biltong supply ahead,' he yelled and there on a sandbank that stretched out into the river sat a crowd of people enjoying sundowners – the classic African sunset beverage.

Once again Ace's family and girlfriend and Reilly's family and girlfriend had come to visit them. This time I too had a family member. My sister Hollie was with them all on the sandbank. *'For some reason this reunion was especially emotional.'* Perhaps it was because *'the constant question was, am I going to come out of here alive?'* With my family all being in Kenya, I had had no contact with them since before we left for the source. At Kariba and in Vic Falls, when the others had had family I had done my own thing, and now suddenly my sister was here and *'I have someone to tell all my adventures to'.*

We spent a fabulous week with family; boating, eating well, game-driving and walking. Mana Pools is honestly one of the most amazing places on the planet. It is wild, it is pristine, there are no people and for a rainy six months of the year it is almost inaccessible.

One morning when I was walking with Hollie and Judy, Reilly's mum, we pushed into a pride of

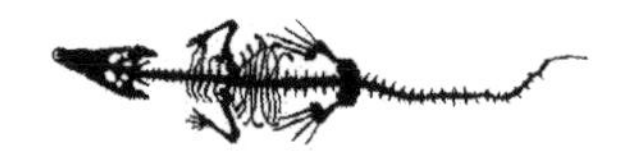

nine lions in amongst the reeds. We stood frozen, three metres from the nearest lioness. Silently we eyed each other.

'Hold your ground. Stand still,' I whispered.

For a full minute we stood stock still until the pride slunk away into privacy.

We sat, Judy, Hollie and I on the banks of the river as we waited for the car to catch up with us and I told them the story of myself and Samuel, my deaf tracker and the charging lion.

I had been working in Makalali in South Africa for some months and was pretty sure of the lay of the land. One weekend we had some guests in who were keen as mustard to see lion.

'No problem' I grinned and mimed to Samuel what they wanted. That night we stayed up talking late into the night and so when my alarm went off at five a.m. I was somewhat drowsy and in prepping my gear forgot to load the gun.

Over steaming mugs of tea we traced roads on the map with our fingers and planned our routes. It was our jobs, a tracker and a guide together, to leave earlier than the guests and to find fresh tracks and thus pinpoint locations of wildlife to save driving times. Off we set into the cold morning and before long Samuel spotted the fresh tracks of a lioness. We radioed back to camp to let them know we were on the trail and then proceeded to follow the tracks. I shouldered my .458 and, Samuel ahead, eyes to the ground, we slowly followed. Before long things started to change, tracks crossed over tracks, skid marks could be visibly seen and we found a few territorial markings. Probably there had been a fight.

We pushed forward ever so slowly, through the scrub, peering around each bush with trepidation. Suddenly a guttural growl sounded and an angry lion, a young male, tore from our right and confronted us.

We froze. This was the first time I had ever been charged. I raised the gun to my shoulder and both Samuel and I took a few steps backwards. The lion growled and came at us, Samuel leapt behind me.

'Fire. Now!' he fiercely whispered.

I released the safety catch and waited. When the lion came within ten metres I fired.

Click. The gun had misfired. I cocked the gun and put my finger to the trigger for the second time as the lion skidded to a halt, its lips raised in a snarl that caught in my throat and gave me a sensation of choking, such was the fear.

Again Samuel and I took a step backwards. No reaction. One more step. Then another. Periodically the lion would start forwards but eventually it let us out of sight and breathlessly we ran to the road where the guests had heard the whole commotion from the car.

'No way!' we told them, 'that lion is out-of-bounds. We'll find another one.' Our knees were still shaking and adrenalin coursing through us.

In the end, with twenty-twenty vision I look back and I am glad that I had forgotten to load the gun otherwise that lion would be dead. Instead we were alive, it was alive and we had a great story to tell!

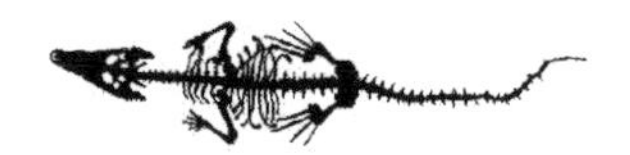

The long days with family and friends on the banks of the river slipped by so quickly and soon we knew we had to push on. It was with refreshed minds and super-fit bodies that we said our goodbyes and pushed on. *'The next time I will see Hollie is at the sea or back at Imire'* I wrote.

John Travers stocked us up with a few six packs of beer and all of the left-over food, and those nights where we were alone in the wilderness with a beer and a golden sunset were exceptional. Nothing beats a cold beer in a truly wild spot.

A day or so after leaving the family we were pulled in by Norman Monks, a ranger with National Parks.

'Young adventurers!' he scoffed and proceeded to give us a severe dress down on the ethics of being seen kayaking down a river with a six-pack of beer each strapped to the front of our boats.

'I want all that beer put away or I'll confiscate it all.'

We took him at his word and stopped to put up hammocks and drink all the beer in one go!

Norman Monks has now been discharged from his post as Zimbabwe goes 'all black.' He is one of the last of his era, a true bush-man who lives and works in remote regions purely for his own love of wildlife.

I thought of Mr. Monks as I wrote in my diary a night after our meeting. *'The most special feeling about this section of the trip is the sharing of the same rights and privileges to the Earth as the animals. I could live out here forever.'*

As always we fished and tracked, relaxed and paddled. Life was good.

Not long after we had paddled over the boundary of Mana Pools National Park, I hooked a green-headed tilapia. Tilapia is one of the species that

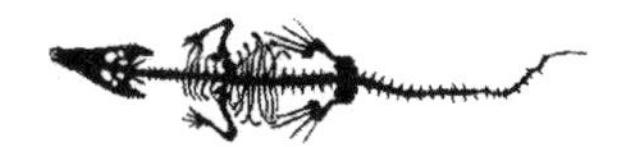

have struggled following the building of the dam. Previously they would nest in the dry season when the water levels were low and the current slower. However the water levels are now very much dam controlled, and the unpredictable (to the fish) water releases often wash away their nests.

The river here had multiple channels, empty of trees except the infrequent *Kigalea* or Sausage Tree, the canopies of which were all of the exact same height at the base – the height to which a large elephant with a stretched up trunk could reach!

Hippos jammed the channels full, and the sandbanks between the polished stone-like hippo backs were barely ten centimetres below the water. In some places the river was so shallow and so full of floating vegetation, dominated by the introduced water hyacinth that we had to disembark and drag our heavy kayaks through the green.

At a lunch stop here, Ace forgot to tie his kayak up and we came back to find that it had floated more than a kilometre downstream!

The waters were so clear and broad that the sky was perfectly reflected and our eyes stung from the UV rays. It was here Reilly managed to get his finger badly bitten by a rabidly angry tigerfish and at night on the grasslands, buffalo came so close that we could hear them breathing as they champed on the short grasses around us.

Before leaving on the expedition, Ace's dad had given us a satellite phone that we were instructed to use only for emergencies, and the occasional call home. Most times we couldn't get the phone to work at all, and when we did it drained all the battery which we then found difficult to re-charge, despite having specially brought along a 'solar blanket' at vast expense. It was this fold away solar charging unit that had given up when Ace had almost drowned in the upped Zambezi. After this, although we kept the phone, it was rarely, if

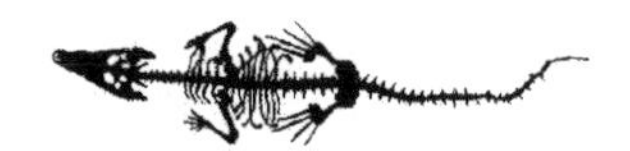

ever switched on. To get messages home we would stop in at some river-side lodge, let them know who we were and what we were doing, and somehow or other a message always reached home that we were alright.

It was a surprise and good timing then that one evening we decided to switch it on and lo and behold a call came through. It was John Travers, Reilly's dad:

'Hi boys, how you doing?' the voice seemed a million miles away, and sitting out there beneath a million stars on the riverbank he could have just been calling off another star.

'Good Dad, we're doing *lekker* no problems. Why you ringing? What's up? Mum alright?'

Whenever there was a call from Imire one had to ask what was up. The political situation in Zimbabwe meant that the Travers' hold on one of the few remaining game reserves was tenuous.

'Yeah it's good Rei. Thing is we have agreed to take on two white rhino and we need to go get them from Matopas. Do you boys want to have a break and come help?'

Although we couldn't hear quite what was being said, Reilly's be-whiskered grin betrayed good news.

'For sure Dad, when you going to get them?'

'Don't know Rei, I'll let you know, but it will be in the next three weeks.' Reilly got further tidings from home and the family and rang off to relay us the news.

James was in a totally different situation altogether. At the family stop in Mana Pools, he had heard that his Dad had bought him and his brother Campbell a birthday present each of a buffalo and lion trophy. This had caused a fair bit of friction amongst us. We were tired from paddling and fractious after a bout of food poisoning, and this news was not exactly a booster.

Our "Row Rhino Row" trip had been talked about from the perspective of conservation. We were to raise the profile of the rhino, white and black, wherever we

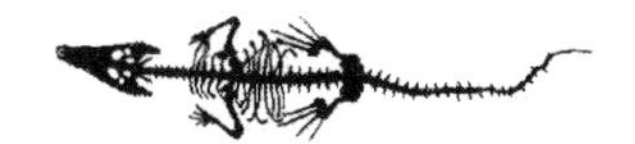

went. Pausing the down-river trip so James could go on a hunting safari was perhaps not such a good move.

Finally after a night filled with the snorting of hippos, we decided to give ourselves a break off the river and time out of reach of the relentless sun. I was feeling slightly off-colour and had been struggling the last four days. We were all tired. If we took a break James could go on his hunting safari with his Dad and brother, and we would head back to Imire and help in the trans-location of two white rhino to Imire Safari Ranch.

We had now reached the area on the lower Zambezi where hunting concessions surrounded us on both the Zambian and Zimbabwean sides of the river. We pulled in the next morning at the camp of Marty Stewart, Mutawatawa. James contacted his family to let him know where they could pick him up and Reilly and I were fortunate enough to be able to catch a lift back to Harare with Alan Hickman, a bush pilot who had just dropped the next set of hunting clients off at the lodge.

It wasn't easy to leave the river at this point. As we stashed the kayaks we wondered if we really would be back. *How long would we be off river for? Would each one of us want to continue? What would happen in the time we were away? Were we going to return and complete this epic journey that we had set ourselves? Was this break an easy way out of something that was sapping our strength, both mentally and physically?*

Our minds were quiet and out bodies tired as we taxied down the dirt landing strip and took to the air, Alan circled the camp. No one of us predicted that it would be a full six weeks before we were to be back together, here at Mutawatawa Camp.

Alan was kind enough to fly low over sections of the river that we had paddled the previous week. It was awe-inspiring to look down at this great meandering length of reflective ribbon filled with ripples from pods of hippos and lazing crocodiles, and to see, through a

wider lens, the trip we were mid-way through undertaking. The decision to have this break had been soul searching and the questions, although un-uttered; lay like hidden elephants below the forest canopy.

We had all lost a considerable amount of weight since embarking on the trip. The hours upon hours of flat, still water in the lazy sections of the Zambezi had tired our bodies and the slower pace of paddling had made the three of us more fractious. Arguments had arisen and then dropped just as quickly.

The paddling was not always easy and I won't deny that there were times when I wished that we had never embarked on such an adventure; times when I wanted to leave the water and the paddling for good. Times when my tired body and mind shrieked at me.

Perhaps a break would do us good.

Part Five

Appendicitis and a rhino-sized interlude

Theoretical conservationists plan for a bright tomorrow;
Active conservationists face a grim today.
Birute M.F Galdikas, Reflections of Eden.

Do I trust you to keep the rhino safe?
And do I trust you to keep the leopard free?
And do I think you'll seal the poacher's fate?
Do I trust you Man? Not me.
Roger Whittaker, Make Way for Man (song lyrics).

Chapter Twelve

Appendicitis

Thou shalt respect thy body and listen to its callings

We landed at Charles Prince Airport just before dusk and Alan invited us to stay with him at his house nearby. That evening we had cold beers and the biggest, richest meal you could think of. It was our first night on a bed and under a roof in three months.

By midday the following day, Rei and I were back in the volunteer house on Imire, recounting our tales midst endless questions. I was still off-colour and, after a beer, quietened down. After a further beer things were not good in the stomach department. Both Hollie and I put it down to the three rich meals I had had since leaving the river and I headed upstairs for an early night. By the time Hollie came to check on me, I was curled in the foetal position and rocking in pain. Hollie rang home to Kenya and Mum, who is a nurse, told her what to watch out for.

'Could be something serious, keep a close eye on him,' she said. 'Let me know in the morning and make sure he keeps drinking.'

At two in the morning Hollie suddenly sat bolt upright and shook me awake.

'Jamie, it's appendicitis! I'm going to ring Judy, we need to get you to the hospital as soon as possible!'

People swam in and out of my consciousness as the stomach pain played leap-frog inside me. Hospital was an hour's drive away in Marondera along bumpy dirt roads, and with low fuel reserves on the farm, fuel would have to be siphoned from other vehicles. We would have to wait until morning to head to hospital.

At six a.m. I was loaded onto the backseat of a pick-up, dosed up on painkillers and moaning audibly. By seven thirty I was at Marondera hospital and suited up ready for an x-ray.

Several times, before we had set out on the expedition, we had been warned that we should have our appendix out, as appendicitis was one of the main afflictions that troubled all long trips. We had glibly ignored all of these suggestions and here I was, three months later, flat out on the operating table having my appendix removed.

'Lucky' the doctor, Kevin Martin, told me. 'A few more days on the river and that appendix would have ruptured and you'd have been one dead kayaker.' He shook his head and put his hands in his pockets. 'You should feel better by late this afternoon. Don't over-do things, try having a short walk,' he motioned over to where a zimmerframe stood in the corner of the small hospital room. 'Maybe go and visit Norman. But mind, only a small whiskey.'

Norman and his wife Gilly were Reilly's grandparents and were iconic in Zimbabwe for their work in running livestock and wildlife on the same ranch. It was they who had built up Imire to the game park status it now held. Reilly turned up later that afternoon and together we ever-so-slowly

walked the hundred metres or so over to Norman and Gilly's house. We joked all the way until my laughter hurt so much that we had to turn serious.

'Reilly man, imagine if we had still been on the river, if we hadn't have gotten to that camp, if Alan hadn't have been there in his plane? You'd have had to be doctor and cut me open with your Leatherman. Do you even know what an appendix looks like?'

'Of course I do Jimbo! Isn't it that long thin thing that hangs low between the balls?'

'Reilly, *punda wewe'*, you donkey, I grinned.

'No Jim, I'd have done it. We had the blades, we had the book, and I'd have followed instructions man. No worries.'

Norman was an old, wise man who shook his head and chortled with us at the whole imagined scenario. With a sparkle in his eye said 'No, Jimbo, you don't want to trust my grandson, he'd have chosen the right Acacia tree to lay you under and that's it.'

Those few afternoons at Norman's were one of the reasons why we survived the next leg of the river. As we sat and sipped on cold whiskeys and ice, Norman regaled us with tales from the old days and coached us on how to fend off croc and hippo attacks, which, he said 'will come hard and fast in the next section, you mark my words.'

He told us that crocs attacked mostly in the summer, and were most dangerous in the breeding season when males fought against each other for mating rights. Female crocodiles were dangerous

while they incubated eggs.

'I'm an old man,' Norman drawled in his old Rhodesian accent, 'and I've looked at your journey and I see what you've done and you've got this far and fantastic for everyone, but –' he took a long suck on his pipe and blew the blue smoke into the hazy warm air '– you guys are actually now going through a time where attacks are very likely.'

He told us stories of people being hit by hippo, their boats up-ending in the water and swimmers taken by crocs. He cautioned us to look warily at all hippos and to judge our paths with care. 'And don't forget the bull-sharks!' he said.

The next three weeks were spent relaxing and recovering at the game park. In my diary I wrote: *'I'm getting fit, strong, recovering, doing a lot of exercise – lots of walking round the game park before I can be fit enough to go and help in the rhino trans-location.'*

Chapter Thirteen

Rhino Trans-location

Thou shalt fight with beast in order only to save them

A single kilo of rhino horn is worth as much as 70,000 US dollars to the end user in Vietnam, sometimes a lot more. It is believed by many to be an elixir for fevers and liver problems, a virility enhancer and a luxury item. It may be gifted to others to show wealth and can be used as a hangover cure after an expensive night of drinking.

In 2009, the year we ran the Zambezi, 122 known rhinos were poached in South Africa. The killing peaked in 2014 at 1,215 animals. 2016 saw a slight drop to 1,054 animals in South Africa alone. This year, 2017 there is speculation that three rhinos a day are being taken! Killed only for their horn.

And what is rhino horn really used for? In 2008 a rumour swept Vietnam that imbibing rhino horn powder can cure cancer. The rumour persists to this very day. Powdered rhino horn is also increasingly seen as a cocaine-like party drug in Vietnam. According to TRAFFIC's Tom Milliken, a 'rhino-horn detox' is the most common routine usage prescribed in the marketplace. Rhino horn is a status symbol. It can grease palms, and it can flaunt wealth in countries from China to Yemen, Vietnam to Thailand. In countries where corruption is rife, you can buy anything with money ... or rhino horn.

In fact, as pure keratin, one may as well just

pulverise nail clippings and sell ... or snort it!

Zimbabwe had introduced a country-wide initiative, led by the well-known wildlife vet, Dr. Chris Foggin, to de-horn every rhino in the country. Part of our trip to Matopas was to help Dr. Foggin in the de-horning of the remaining rhino, while at the same time sourcing a suitable young male and female white rhino, which we could trans-locate to Imire.

It was Norman's dream to establish a breeding herd of white rhino on Imire. Decades before, this region had been the haunt of these magnificent beasts until they had all been shot out in the mid 1900's.

'Reilly man, have you packed your Leatherman?'

'Yebo'

'Food? Water? Fuel?'

'Yebo'

'Toolkit?'

And we were off. Jon Olivey, Reilly and I.

'Hey Jamie,' Rei teased, 'you better watch that M99, don't go drinking it!' Rei was referring to a somewhat embarrassing incident that had happened sometime before at Imire. An incident that involved myself, a girl and rhino sedative.

Reilly has a cousin who at the time, I was very fond of. One evening at Imire we were to have a party to which both she and a very good mate,

Kitkat, were invited. Now Kitkat, Rei and I had the international mates deal: Friends before women. Some months previously I had been in Mozambique and had bought a BMW 650 motorbike of which I was exceedingly pleased and proud. Several times I had taken Rei's cousin out for a spin and she loved it. I came back that afternoon to the volunteer house on Imire with plans to take her for sundowners before the party began – only Kitkat, my motorbike and Tara were missing!

I was furious and dragged Rei to the bar.

'Rei I'm putting a drop of rhino strength sedative in this beer. Don't drink it OK! It's for Kitkat.'

Rei could see I was fuming and tried to calm me down. 'Jamie I don't think that's a good idea. What will we do when he goes out?'

'I don't care Rei. There is that truck that is leaving to Mozambique tomorrow – we can put him in that and he will wake up in Mozambique with a good story to tell his grandkids!'

Rei finally agreed and we carefully placed the beer where Kitkat would be sure to reach for it. As dusk fell and the party got underway the beer must have been moved and somehow I realised I was drinking it after only a few sips. I retreated to the sofa and reclined as my vision began to blur.

'Rei' I slurred, 'where are Kitkat and Tara?'

'At the bar,' Rei replied. 'Don't worry Jamie nothing is happening.'

Reilly took over the telling of the story.

'Crikey,' he said as we rattled over corrugations on our way to Matopas National Park and the rhino trans-location. 'Jimbo decided that he had had enough and was going to ride back to Kenya on his motorbike that night. We thought he was just messing but anyway, come morning, there was no Jamie and no BMW.'

Olivey broke in to Rei's story 'Jamie only you – idiot!'

'Yeah well we told Mum, and when we couldn't reach Jimbo's phone we thought we better ring ahead and warn his folks in Kenya where he was. Then at about three o'clock that afternoon we get a message that he is on his way back. So he arrives on dusk, tail between his legs and more than a little hung over.'

I took up the story again, 'I'd headed off to Kenya and gone in the wrong bloody direction. I woke up to find myself chained to my motorbike which was chained to a tree on the side of the road (a usual method I used to stop theft if I had to stop to nap). I had no idea where I was and flagged down a car who said I was four hundred kilometres from Vic Falls!'

Olivey and Rei roared with laughter and poked fun at me for the rest of the drive to Matopas. When we pulled in Dr. Foggin and the chopper pilot, Barney O'Hara, were already on site together with the hired staff. The sun glanced off the dry browns of the water-parched landscape though the morning air was still cool. Preparations began almost immediately. It is incredible *'what you need to catch a rhino and de-horn it. We had to sort out the lorry, we*

loaded the crates, there are lots of chains involved, lots of heavy work: there are punctures, engines need to be fixed – there's ... it's just a nightmare - the whole operation is just full of big heavy machinery, cranes, ropes, helicopters, aeroplane fuel – man we were tired at the end of the preparation time'

Dr. Foggin briefed the team. 'Right boys; check your radios and your GPS. Any one of these go out, you let me know immediately. Jamie, Reilly, and Olivey you'll come in the chopper with me and you'll do exactly as I say, no erring, got it?'

'Yebo,' we all three said right on cue.

'Spotter plane is up now, sightings will be GPS'd and radioed in -' the radio at that moment staccatoed across the bush.

'White rhino, mother and calf at 33° 20' 86' ' S and 30° 31' 31' ' E, over.'

Dr. Foggin answered without a moment's hesitation, 'GPS reading and point taken, over and out.'

'Jamie, Reilly, Olivey - in' he gestured to where the chopper blades were already spinning and the dust whirled up in choking clouds. 'The rest of you' he shouted over the roar of the noise, 'listen in! When we call you, you get there as soon as possible.' The others nodded. Question time was over.

There were five of us in the chopper including the pilot. Rei, Dr. Foggin, Olivey and myself. The doors on each side had been removed

so as to make for better sightings and in a whirl of dust we lifted into the cool morning air. We zeroed in on the reported sighting and were soon hovering over the frantic rhino and her calf.

In any darting operation the pilot and darter have to be in total balance with each other. Each needs to be able to read the other's mind; the darter needs to be able to plan his shot to the exact movements of the chopper. Often the pilot and darter have worked as a team together on many operations and will work with no-one else.

At a flick of Dr. Foggin's fingers, Barney lowered the chopper within firing range, and Dr. Foggin fired and reloaded another dart from the three he had ready.

'Take two,' he muttered under his breath, aimed and fired and there, down on the bushveld were two rhinos running towards cover with two crimson darts protruding from their rumps.

'Stopwatch' Foggin yelled.

'On it!' Olivey yelled back but Foggin was not listening.

The spotter plane lay to our west and another radio call came in 'located and watching' the pilot hummed.

'Four minutes,' Olivey called out.

We each kept a close eye on the rhino below. A few minutes ticked by till the M99 kicked in. Already they were showing signs of slowing and faltering.

'Seven minutes.'

Exactly on cue, the rhino collapsed to the ground and Barney was bringing his bird in for landing. As the skids touched down, we were already off-loading: ropes, chainsaw, machetes, water and spray gun. Foggin was already at the mother's side taking a blood sample while Rei cleared the surrounding bush so that we could roll the rhino into a stabilised position.

'Rei, get over here, left hind leg, OK roll!' We rolled the rhino onto her front, legs tucked under her where possible. Foggin quickly covered her eyes and stuffed cotton wool into her ears. Within minutes the calf was also in the same position, just metres away. We heaped fresh, green brush over their heads which we then sprayed with water. We sprayed their bodies too and Foggin kept a close eye on their blood pressure as he ear-notched and tagged the rhinos with a bright yellow numbered tag that would help monitor them. Then he started the small chainsaw; pulling down his safety goggles he cut into the horn an inch or so above the nose.

Major blood vessels run through this section and so even de-horning the rhino leaves an inch or so of horn that poachers still kill for. With the price of powdered rhino horn so high, even this inch could make someone rich.

As each horn came away, it was placed in a hessian sack and reloaded into the chopper along with all of the scrapings that were picked up too. A tiny tracker chip was inserted into the remaining horn and the antidote, M5050 was administered; within minutes the rhinos were coming round. They were sprayed once more, the brush and eye-covers

removed and we all stood back. But there was no time for us all to watch. While Foggin watched, we re-fuelled and cleaned the chainsaw, reloaded the chopper and re-filled the water spray canister.

As the rhinos stumbled to their feet, the spotter plane was already radioing in new co-ordinates.

'White rhino male, 33° 28′ 31′′ S and 30° 45′ 36′′ E.'

We piled into the chopper and took off again; the day was heating up, the temperature would soon reach thirty degrees centigrade, a dangerous heat for administering drugs. Below us the mother had made it to her calf's side and the pair was groggily heading to the deep shade of some acacia scrub.

'They'll be good,' Foggin mouthed. 'Thirty-nine more to go.' We had four flying days in which to complete the task and we knew that with poaching at such high levels we were unlikely to reach this quoted figure. Sure enough, within minutes Olivey pointed to our right.

'Carcasses!' he yelled. 'Two,' and there below us lay two old rhino carcasses each with their horns messily hacked away with what looked like machetes. Already some of the bones had been stripped of their meat by hyena or jackal and had been bleached by the sun. Around the bones grazed a herd of impala and a single, darker coloured tsessebe.

We touched down near the darted male rhino. Its size was awe-inspiring. Three tonnes of battering ram. As we leapt down from the chopper,

Olivey made the almost fatal mistake of running beneath the tail arch.

'Olivey, no!' Reilly screamed a warning. The rotor blade was still turning despite the engine having been switched off. Olivey raised his head and looked towards Reilly and me. Both of us had stopped breathing. This was the end, this was it! Everything turned to slow motion and as the blade came down towards Olivey, he stepped forwards and the rotor caught the billowing cotton of his shirt. The rip sounded loudly and seemed to surprise Olivey who had no notion that he had just escaped a severe accident.

'Idiot!' Barney made a chopping motion across his throat with his hand. 'Do that again and you are a dead man.'

Olivey had no answer, he didn't need one, the de-horning was already in progress. Over the next three flying days we spotted numerous carcasses and managed to dart only nine white rhino and one black. The census had to be re-adjusted. We had lost twenty nine rhino to poachers if we assumed we had not missed any other live spottings. We almost lost the one and only black rhino too. The spotter plane had radioed in his location.

'Black rhino male, tough territory 33° 53' 43'' S and 30° 28' 47'' E

Barney wasn't so sure about landing his chopper. The rhino was still standing but we were on rocky ground, the rocks being part of the lower slopes of a massive rock kopje. Away from us we could see a lone buffalo male trying to move quickly down the rocky slope to escape the noisy bird above

him. Finally Barney spotted a shelf of sloping rock where we might just be able to land but Foggin called him to circle again; the rhino was faltering and losing his footing. As he fell, he rolled down the rocky slope and lodged, upside down in a crevice.

'He's down! This is a tough one boys, he might be a goner.'

We landed and squeezed carefully away from the chopper blades and rushed down the crumbling slope to his side.

'Fuck, we're gonna need more people,' I yelled. Foggin signalled to the chopper pilot, 'fly back - we need more support. Quickly.' Barney took off and precious minutes sped by as we waited for his return. Foggin radioed the ground crew.

'Four of you are on standby. Chopper coming in now, load more ropes and get here quickly.'

'Affirmative!' someone radioed back.

With the four of us in place, Barney landed his bird with the four new recruits and all hastened over to help. Even now the rhino's breathing was becoming laboured.

'Three people here,' Foggin signalled to the rear end, 'and the rest of you this side. Thread the ropes under him and let's lift the front end first and then you guys at the back, as his shoulders lift free, push and try roll him out of the crack onto his left side.'

As he spoke he was fitting a cloth mask over the rhino's eyes. His ears already held cotton wool. It took a good five minutes to perform this

manoeuvre and only adrenalin gave us strength to lift his dead weight against the slope of the hill.

Finally we were back on target, a fine mist of cool water spray dampened his tough hide and a wind picked up that helped to cool him further. Foggin took a blood sample and tagged and notched his ears. He affixed the GPS tracker so as we could always find him. We pushed back to the chopper and watched closely as he revived, and we collectively sighed as twenty minutes later he was on his feet and headed down the slope away from us.

'Thank God' Foggin whispered.

Foggin is easily one of the best wildlife vets in his field. He has been practising for over twenty-five years and in that time has lost few rhino. He has continued to work tirelessly in his self assigned task to preserve Zimbabwe's remaining wildlife and against the odds to save the few remaining rhino. Recently he moved to work for the Wildlife Veterinary Unit where he now has to deal with mostly corruption fuelled by a desired ignorance and incompetence of a once world renowned Zimbabwe national parks service.

Barney had to do two short chopper flights to get us back to base, and with cold sodas in our hands, Foggin handed the sacks of rhino horn over to the head of Zim Parks, a skinny man with an over-tight belt and sweating face who grinned as the sacks were transferred to him. Months later we learnt that the park safes had been robbed and all rhino horn had disappeared.

Now only two darting operations remained, and these were the two rhino that were to travel to

Imire, sixteen hour's drive away. Foggin had already darted and tagged the two that had been selected and had fitted them with GPS trackers that meant they could easily be found. As it turned out this operation would be a whole lot more difficult.

Early morning on the following day Reilly, Olivey and I again hopped into the chopper and we took off with Foggin and Barney at the joystick. Barney flew low over a journey of cantering giraffe that streaked over the ground below, dodging between scrubby thorn trees and green, algae like pools, dry of water. Foggin tracked the young female rhino first and in the thirteen minutes it took for her to go down; we had radioed co-ordinates to the ground crew who were on their way with the massive single crate truck and cranes. Luckily she had gone down relatively near a road which made the operation slightly easier. The Land-Rover was ahead of the truck and a team of sweating, lithe bodies frantically cleared a road in front of it as they headed towards us. The newly hacked road had to be big enough to allow a double-axle truck clear access to the downed rhino. Once they reached the rhino, the Landy turned around and with three people on the roof; any overhanging trees were hacked away.

Meanwhile our chopper team worked on keeping the rhino cool with water spray and monitoring all vital signs for change. Finally after what seemed a helluva long wait, the truck could get in. Immediately the frenetic activity flicked up a notch.

'Crane operational' yelled the crane driver. Someone had scrambled to the top of the crate.

'Lower her, bit lower, halt!' He hooked the crane hook onto the ring from which four strong chains snaked to each corner of the old shipping crate where they were tightly secured.

'OK take her up.'

The crane took in the slack and slowly the chains tightened and the crate began to rise and to twist in the air.

'Take her left!' The crane arm slowly moved the crate away from the bed of the truck and lowered it to within grasp of the men with their hands raised.

'OK let's twist her clockwise,' Foggin yelled as he glanced across at the rhino. It had been nearly a full hour now. The men worked together to twist the crate so that its door was positioned towards the front of the rhino.

'Alright, positions ready – lower her!' The crane slowly lowered the crate as each man held his corner steady. A small puff of dust blew up as the crate touched ground.

'Alright, man the door.' Foggin needed two men up on the crate so that as soon as the groggy rhino had entered the crate, the sliding door could be let down behind her. Foggin came to where we stood looking down at the massive body of the rhino,

'OK boys, get behind her, don't be aggressive but use your strength. Right, drug-reversal in –three minutes, two minutes, one minute, ten – nine – eight.' The massive female started to twitch her ears and began to attempt to stand up. 'Seven –six –five,'

she stumbled to her feet and shook her head, the eye mask still in place. 'Four – three' she took a tentative step forwards. 'Right, hands on her ass, two behind; one each side – two – one and push'

We each put our weight into it and gently eased the unknowing rhino forwards into the open crate; the door slid closed between us and her massive ass. A fine mist of water was sprayed over her from above and we waited ten minutes or so for her to get her balance and for the M99 to wear off before the crane clankingly lifted the crate back onto the low loader. Foggin checked and double checked that the crate was secured and then signalled for the convoy to leave. The truck bumped happily behind the Landy as we climbed back into the chopper for the final time.

The male that we darted went down much further from the road which meant that the truck took a lot longer to reach us. Foggin had to administer a second dose of M99 and it was a full four hours before he could be loaded into the crate and the truck could start its journey back to base camp.

By now it was almost midday and the last straps were being secured on the two crates that each contained one rhino and stood, one behind the other, on a large twelve-wheeled low loader. The final pieces were loaded and the truck driver signalled that he was ready to go.

'Remember,' Foggin yelled as we began to pull away, 'those rhino must not go down. Keep them standing the entire way. Catch!' he yelled as he threw a 1.5 litre bottle of ice cold water, one at me and one at Reilly. Reilly and I would undertake

the sixteen hour journey sitting atop the crates with some of the hired staff. It was our job to monitor the rhino, to keep them cool and above all to make sure that they remained standing throughout.

It was going to be a long ride home.

Travel route from Matopas to Imirie

Reilly grinned across at me 'Comfortable Bru?'

I smirked in reply. *Comfortable?* There was no way this was going to be a comfortable journey. I swallowed half of my cold water in a single glug, Reilly did the same and we settled in for the journey ahead. Half-hourly we scampered over the top bars of the crate in an attempt to ease our aching rear ends.

'Rei, did Foggin tell you the story of the last trans-location?' I asked.

'No, tell me, what happened?'

'One of the rhinos managed to break the side of the crate open and the opening was big enough for her to put her head out, and they had to go through Gweru with this bloody rhino's head stuck out the side and all the locals cheering and yelling which made her even more nervous and feisty.'

Reilly grinned at the thought of it. Most of Zimbabwe's locals had never seen a rhino, let alone begun to think about conservation of this iconic species. Our "Row Rhino Row" expedition had been one way to bring news of the demise of this great beast to the people who could save it. Until conservation worked from the bottom up, we were in for a rough ride. All over Africa, the advent of the Chinese and the sheer market demand had put many species in jeopardy of extinction. The Chinese are intricately involved in wildlife crime, from being a hub of trade to providing market demand.

Time and time again one or the other rhino would go down on its knees and we had to jump

down and coax it to its feet using rope and even a low voltage cattle-prod, all the while being extremely careful not to let the rhino know that it could easily trample us. The minor travel sedative helped to keep them calm and slightly woozy so they stayed alert without reacting to every sound and smell which would have been dangerous and potentially lethal.

As dusk drew in, we stopped for a while on a quiet bit of roadside so we could re-stock with water, take a leak and give the rhino some quiet time. Twenty minutes later we were back on the road. The dark closed in and soon it became hard to see anything around us. We had a torch each so as we could see the rhino but the flimsy beam could not reach very far ahead of us. Sometimes we managed to catch the amber eyes of a bush-baby in the roadside trees. I moved to the front of the crate and snuggled deep into my fleece and barber jacket (an oilskin, as they were once called).

The night air was cold and Reilly and I had no more stories to tell. We were exhausted. Reilly sat at the back of the crate and now and again I could hear him faintly talking to the rhino, more to keep himself awake than to soothe the massive beasts.

Suddenly without warning something struck me hard against the bridge of my nose. Reilly was whipped out of a semi-sleep when he heard the crack of bone across the noise of the engine. The blow was hard enough to momentarily knock me unconscious and when I came to, seconds later, still seeing stars, I found myself in the crate with the rhino.

The situation was now hit-and-miss; I didn't

have the strength to get myself out from where I had fallen, nor could I reach the sides of the crate. The noise of the engine meant that even calling for the driver to stop would be futile. Although he couldn't quite make out what had happened, Reilly reacted quickly.

'I'm down Reilly, I'm down, mayday!' I called weakly, still heavily stunned.

Reilly was in the crate with me already, standing on the rhinos back. It was the torch beam that saved me. If it hadn't been for the torch, Reilly wouldn't have known where to look. It was pitch-black in that crate.

'Here Jimbo, grab my hand!' Reilly shone the torch into my eyes. I could feel blood trickling down into my mouth and together the blinding glare of the torch and my swimming head meant that I couldn't quite grasp what Reilly had said.

'Reilly you're blinding me, I can't see.' The rhino was getting aggravated.

'Jimbo you need to take my hand.' Rei moved the torch beam to my left and I saw his hand and grabbed it with all my strength. Reilly pulled hard and I managed to gain footing on the rhino's back. Reilly shouldered me up towards the top of the crate and I managed to pull myself over the edge, Reilly right behind me. We quickly dropped down to the side and Rei kept a close eye on the now fractious rhino through cracks in the crate while I tried to clean the blood away from my throbbing nose. It appeared that I had been struck by a cross road cable that, had it been a few inches lower, it would most likely have decapitated me. A thought I did not

dwell on.

Finally at four a.m. we pulled up to the newly-built bomas on Imire. John, Judy, Hollie and the team of volunteers were there and ready to help. They had been expecting us many hours before and Foggin, who was in the Land-Rover ahead, pulled up twenty minutes ahead of us in the truck to find them all asleep on the gigantic bed of hay that had been set up for the arrival of the rhino. Headlights swerved one way and then the other across the scene, while Judy poured herbal concoctions at all the corners of the enclosure.

From her youth Judy, and in fact most of her family, had always been fascinated by alternative medicines. Whenever we were staying with Judy or with her sisters, we would be regaled with intriguing tales of herbs or actions that had worked.

'I promise' Judy would swear on whichever wild animal was beneath the table at the time, 'I swear on Pog's (the tame warthog on Imire), that the bat guano made his hair grow back!'

But now, if the herbs were going to work then we wanted them. We wanted anything that would make this rhino business flow easily.

The water trough was filled with fresh sweet water and fresh browse and Lucerne was laid out in anticipation. The massive truck was worked from all angles ... forwards, backwards, right hand down, forwards a bit – OK backwards with left hand down ... as it reversed up to the stockpile gate.

First one crate and then the other was lowered to the ground, the upward sliding doors

opened and slowly each rhino was prodded and pushed by Foggin from above, until each moved forwards and into the boma.

Olivey and Hollie worked together to ensure that each and every log was secured at the entrance and then we all backed off. The truck was taken out onto the main farm road, the Landy headed back to the main farmhouse and we piled in with the rest of the volunteers to head back to Numwa House for a much needed sleep.

Only Judy, John and the tireless Foggin stayed behind. With them was a young 'rhino translocation specialist,' not veterinary related, but there to learn on the job. They would constantly monitor the rhino over the next two weeks before they were released.

Chapter Fourteen

Two Rhinos Escape

Thou shalt pursue the rhino until they are saved from extinction

By noon two days later the rhino had still not eaten or taken any water that anyone had seen. Things began to get heated.

'If they don't drink or eat by the end of the third day,' Foggin had instructed 'then the situation is dire. Monitor them closely for the next four hours and if they don't drink you'll have to do an early release.'

Foggin was long gone on another trans-location exercise and our rhino specialist was a bit green behind the ears. John took the decision some twelve hours later.

'We gotta let them out boys. Get your horses, station men all around the Chiwawi boundary and make sure all the radios are in use.'

Sixty minutes later the rhinos had been released. Ten minutes after that the mayhem started. Imire is spilt at several points by an all-access main road which breaks the farm into four distinct regions: Chiwawi, where the rhino boma was, Welton which lay across the public access route from Chiwawi and Numwa which bordered Welton. Finally there was the main farm section that was separated from Numwa by the tarred road.

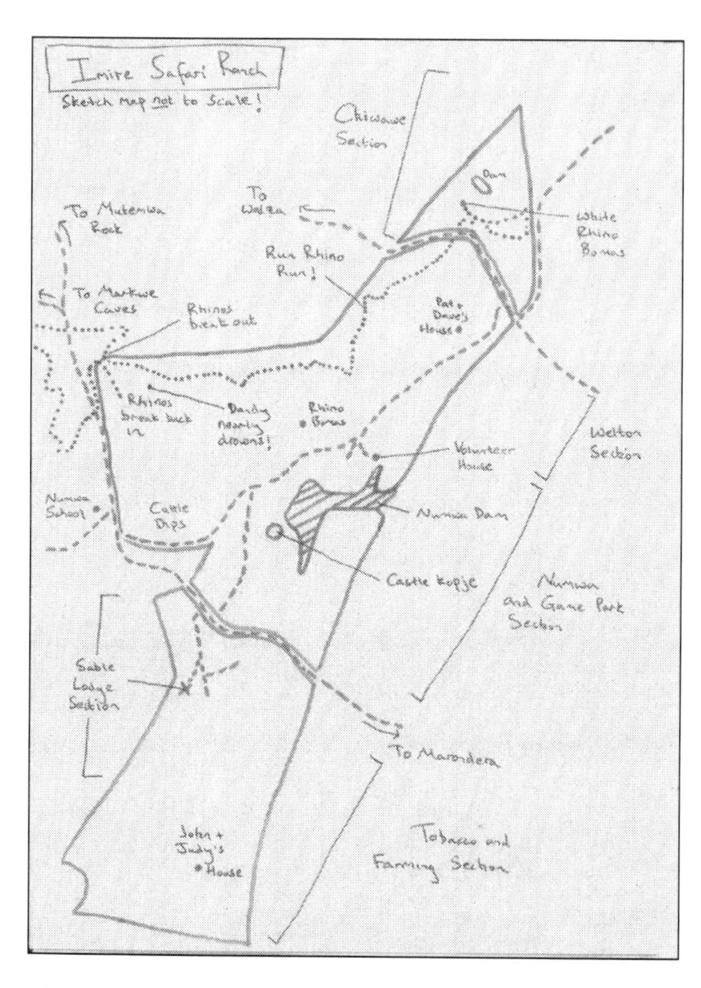

Sketch Map showing the lay of Imirie Safari Ranch

Water had been placed at points all around the Chiwawi boundary fence and, at Judy's insistence, fresh rhino poo too. As the rhino were released they made a beeline straight to the boundary of the Chiwawi/Welton section between which ran the road.

We waited - would the fence stop the charging rhino? It didn't!

They smashed through it as if it was mere branches, over the road and then straight through the Welton fence, snapping the wire as they went. These rhino were on a mission and "Row Rhino Row" quickly became "Run Rhino Run" as a radio call went out across the farm.

Reilly and I had saddled up and spurred our horses, Drummer and Dandy on as we tried to keep up with the racing rhino runaways. The rhino streaked through the grass, crashed through each and every fence onwards into Numwa section. At this point they could either branch left and possibly break into the main section of the farm or they could continue on their chosen bearing and break through the boundary fence and into rural community land.

'Now hell-for-lather broke loose. Two brand new white rhino were now heading for the hills, up into densely populated lands in Chief Soswe's territory. So it was all or nothing. Man, we would have to get these things back.'

The next few hours were pandemonium as radio messages contradicted each other and cars, horses and motorbikes ran in circles round each other across the land. At six p.m. John Travers

called in the troops.

'Everyone with a radio listen up, we can't chase these rhino any more today or we will kill them from exhaustion. Come back to main house, get yourself washed and drink all you can. In an hour and a half from now we will convene for a team talk. Over and out.'

The sun set that evening in a blaze of liquid amber that we watched with heavy hearts.

All search teams had obeyed John and now we were all sitting in the Travers' garden with pints of iced water with freshly squeezed lemons, having had a long swim and wash in the dam. The sun had gone now and our shadows fell across the lawn. The mood intensified as John took his post.

'Right everyone,' he said, 'Reilly, Jimbo, Olivey you guys brought these rhinos here, it's your job to get them back.' We nodded solemnly 'Tomorrow before dawn I want you to take the position that you left off today. Let's assume that the rhino are exhausted and are going to stop to rest. I'll get out on the micro-lite and spot them as the sun rises. Then Jamie, Reilly, and Olivey you get in front of them and round them off.'

The plan was fine-tuned here and there and then we all turned in for an early night. Tomorrow we would need all the strength we could muster.

We woke the next morning before dawn and by ten a.m. John had pinpointed the new position of the two rhino. Olivey was managing the lodge and he had been willingly roped into helping, his light

humour and his quizzical questions keeping us entertained as we waited. Finally John shouted instructions on the radio and the rhino tracking began in earnest, every man following his route as best he could.

Olivey headed towards the two rhino in the pick-up while Rei and I were saddled and ready to catch their trail. Once more the two rhino were fast on the move. My appendicitis scar ached like buggery and our asses were saddle-sore.

Finally Rei and I managed to get in front of the rhino and to head them off slightly. We were upwind of the rhino and the smell of sweaty horse stalled them slightly giving us the advantage we needed. Rei curved around them in a pincer movement and then we yelled and shouted and clapped to turn them back towards Imire.

Would they turn?

We waited with baited breath. Olivey queried our techniques over the radio and the crackle of his voice seemed loud and harsh. The rhino paused, only going through the motions of grazing as it served to calm them. They were distracted. Minutes ticked by and finally the rhino turned back towards Imire.

It had worked; we had sealed the deal; now we only had to herd them across the fence-line.

'Rei, keep her turned' I shouted as the female made as if to break away. The rhino were tired and not moving so fast now, but our horses, Drummer and Dandy were tiring too and it wouldn't be long before they collapsed. We needed

reinforcements.

'Olivey man, we're coming up to the Numwa school road, where are you?'

'Roger that, I'm on that road, one k away.'

'Who's with you?'

'Crispin's on the motorbike. He is right behind me.' Crispin was Imire's head of security.

The road came into view and farm pick-up with it. 'I can see you,' Olivey said. 'Where are the rhinos crossing?'

'Right in front of you – right - now.' I watched them cross and watched Olivey and Crispen take up the homeward chase.

The next palaver that broke over the radio was where to chop the fence for the rhino to re-enter the game park. Each time that they chopped the wire the rhinos changed direction until the fence had been chopped in four or five different places. Finally at about three that afternoon the two white rhino re-entered the game park. I think everyone loudly breathed a sigh of relief. I certainly did!

The local community meanwhile had been avidly watching the chaos, and I'm sure wondering what all the fuss over these animals was all about. '*Hmm!*' we thought, '*lots of people ...*' *It* was at this point we made use of the masses and their pots, pans, empty buckets and sticks. We lined them up along the northern boundary and for over half an hour they sang and shouted and beat their pots and pans. We hoped that this ruckus would be enough to stop the rhino coming back to this boundary any

time soon.

Rei and I were exhausted. Drummer and Dandy needed to drink and we wanted to cool off, so we headed down to a grassed pool on the Numwa River that lay not far away from the northern boundary. The pool was deceptive as it was very deep in the middle and it was easy to get out of your depth. When we got there, both Rei and I stripped off and we leapt butt-naked into the pool to cool off.

Putting out shorts back on we first led Reilly's horse down to the water to drink. There was only a single flat point where the horses could access the water as the banks were so steep. As Dandy slaked his thirst, we poured water over his sweating body with cupped hands. We then led Drummer down. I'm not sure exactly what happened but suddenly Drummer slipped on the smooth rocky shelf and fell into the deep chasm in the middle of the stream. He completely submerged, and as his head came up he was kicking like hell.

'Reilly man, grab the halter!' Reilly dived into the water and came up on Drummer's left side grabbing the halter as he did so. Thrashing about in cold water with an exhausted horse is not something I want to repeat. For several minutes we frantically tried to get Drummer out of the river but the bank where he had slipped in was too steep, and in his exhaustion he was not working to help us.

'Jimbo, this isn't working, I can't hold him up much longer.' It was taking both of us just to keep his head above the water.

'Hold on Rei, I have an idea. You've got to

hold Drummer up.' I swam to the bank and climbed out. 'Pass me his reins.' Reilly coaxed the horse to the bank and handed me the reins which he had flicked over Drummer's head. I tied Dandy to a tree and then tied the reins of Drummer to his. We would use Dandy in the rescue.

'OK now give me his two front feet.' Rei and I heaved and pushed until we had manoeuvred the two front legs of the horse straight out onto the rock, 'OK hold him there. I dived back in. 'Right now we both got to get under his bum and push like mother f####rs'.

'On three then,' Reilly said. 'One – two-three.' We both took a deep breath and swam beneath Drummer's bum: Each supporting him on a shoulder, we pushed off the bottom that we could now reach, and with a strength borne only out of adrenalin, we somehow managed to lift him enough that he could get a foothold.

'Once more, Rei, this time it's just you.' As Rei dived under, I whipped Dandy with a long stick to make him move forwards. As he edged forwards he exerted the tension that we needed on the reins to which Drummer was tied, and with a lot of scrambling and slipping Drummer made it up onto the bank, clear of the water. We jumped out and helped him up, leading him well away. His eyes rolled and his mouth frothed but he was safe. Holy be-Jesus, kayaking was a doddle compared to this rhino trans-location!

We led our pony's home at their own pace, which chimed well with our own exhausted bodies, we left them with Judy where they were given three whole weeks of R&R with good food and plenty of

TLC and a few spicy treats before they were declared fully recovered.

That night we went to sleep knowing that the rhino were back in the game park. We still have a rhino-sized journey ahead of us, but I know I have it in me. We now need to mentally prepare ourselves and to think about all that needs to be done for our return to the river.'

The six weeks certainly had been eventful, but now it was time to head back to the Zambezi. Ace was back in the hunting camp and had let us know that all our kayaks and equipment were still there and safe. Rei and I headed to Harare where we had organised with Alan Hickman (the bush pilot who had flown us off the river), to catch a lift back into the camp. Plans were to head out straight away but Alan delayed a day and so we decided to have a leaving do at a place called 'Bucket Bar.' Reilly had a funeral wake to go to in the afternoon/early evening so we agreed to meet up afterwards.

I got to the bar about nine p.m. and there was no Reilly. I wasn't too worried and ordered a bucket of drinks. Olivey and a load of other mates were there and we started drinking. Bucket Bar was a uniquely Zimbabwean bar. Due to the economic downturn of the economy and the introduction of the US dollar in Zim, prices had risen considerably. It was incredibly difficult to get booze and several enterprising people were smuggling in vast amounts from Mozambique or South Africa.

Bucket Bar was open on weekends

only, and in order not to be raided by the police, and caught with huge amounts of booze, they made their patrons take the risks. The minimum number of drinks one person could order was six. Six bottles of beer, six bottles of vodka, six bottles of wine and you had to pay in USD. As six was generally too many for one person to carry or to drink, the bottles were placed in colourful plastic buckets to which a bottle opener was attached. Ten metres from the bar was a swimming pool with bench-tables and recliners; if there was a bust we could claim it was a perfectly legal house party!

Finally, I ended up leaving. Matt lived near-by and I circled the plot blindly looking for his house; while Olivey, who was going to the same house and who left slightly before me had apparently circled the block the opposite way. We met on a dark corner.

'Who's that?' I have to admit, I was slightly jittery. Where was Reilly I wondered?

'It's me.'

'Who's me?'

'Oh ... Olivey'

'Listen, Olivey' I slurred, 'I've been round the block once and Matt's house is not there, we must be on the wrong bloody block'. We circled the next block and then the next till we found Matt's house and our beds. In the morning we woke to Reilly hammering on the door.

'Let me in boys, let me in. Listen you plonkers, wake up.'

Reilly was a mess, black eyes, cut and bruised and the smell of raw sewage wafted from him. I held my nose in my cupped hands and tried not to gag. A hangover and sewage!

'What the hell happened man?'

'Mugabe's police,' he spat. 'After the funeral yesterday, Candice, I and two other girls, we took the wrong turn and you know that road that they don't let you down after dark, the one that goes past Mugabe's place?'

'Yeah'

'Well - we went down there.' He stopped to ask if he could have a mug of tea.

'Thugs pulled us over and held me face down on the car bonnet and held a gun at my temple. Man - they wouldn't let me tell them that I'd taken a wrong turn and they kicked me about, punched me and all the time I was thinking *shit, what about the girls*. Then they told the girls to wait and they took me into some dude's garden. There were flowers and everything,' he paused to scratch his nether regions, 'Jamie I even saw that *Afzalia quansensis* that we had been looking for.'

'What! You even noticed a tree when these thugs were beating you up? What happened then?'

'They pushed me into, like this sewer, and made me swim laps of this thing. Can't you smell it?'

I wrinkled my nose. 'Of course you idiot, of course I can... '

'Man, Jimbo, I didn't think we were going to

get out of that one. Then they hauled me back to the car and told me to leave.'

'Ay yai yai yai,' I breathed, 'you were pretty lucky.'

Reilly nodded and downed his hot mug of sugared tea in moments. I think he was still reeling in shock.

'What about the girls?'

'They just stayed in the car and when I finally got back we left pretty smartly. What else were they to do – they had to wait. They were pretty damn worried I'll tell you that.'

Rei showered and then told the story again for Olivey, ending as Alan called and said we should head to the airstrip. By midday we were in the air and heading high over Harare back to the Zambezi, I lost in my thoughts and Rei lost in cuts and bruises from the night before. There was a very real chance that infection might set in and so we decided to take an extra load of *muti* or medical supplies with us, as well as staples like rice, salt, sugar, muesli, powered milk and tea, biltong and chocolate.

By evening we were with Ace having beers on the river, recounting our stories and discussing the final leg of paddling. My parents had just told me that they were arriving at the end of November for three weeks so I now had a deadline to work to. The question was, would we make it?

'We have a new lease of life, a new tone of living, a new set of stories to tell each other down the river and I think we are going to make it. Three old friends back

together' is what my diary entry for that night says.

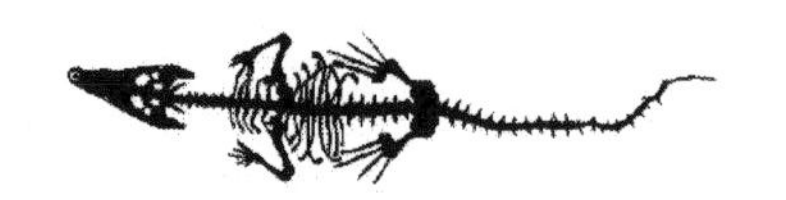

Part Six

An End to the River's Rhythm

Maybe that is why you seem to live more vividly in Africa. The drama of life there is amplified by its constant proximity to death.

Peter Godwin, When a Crocodile Eats the Sun: A Memoir of Africa.

Sometimes luck is with you, and sometimes not, but the important thing is to take the dare. Those who climb mountains or raft rivers understand this.

David Brower.

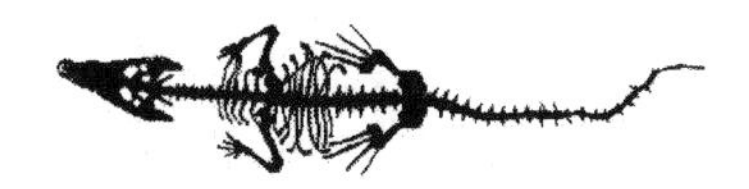

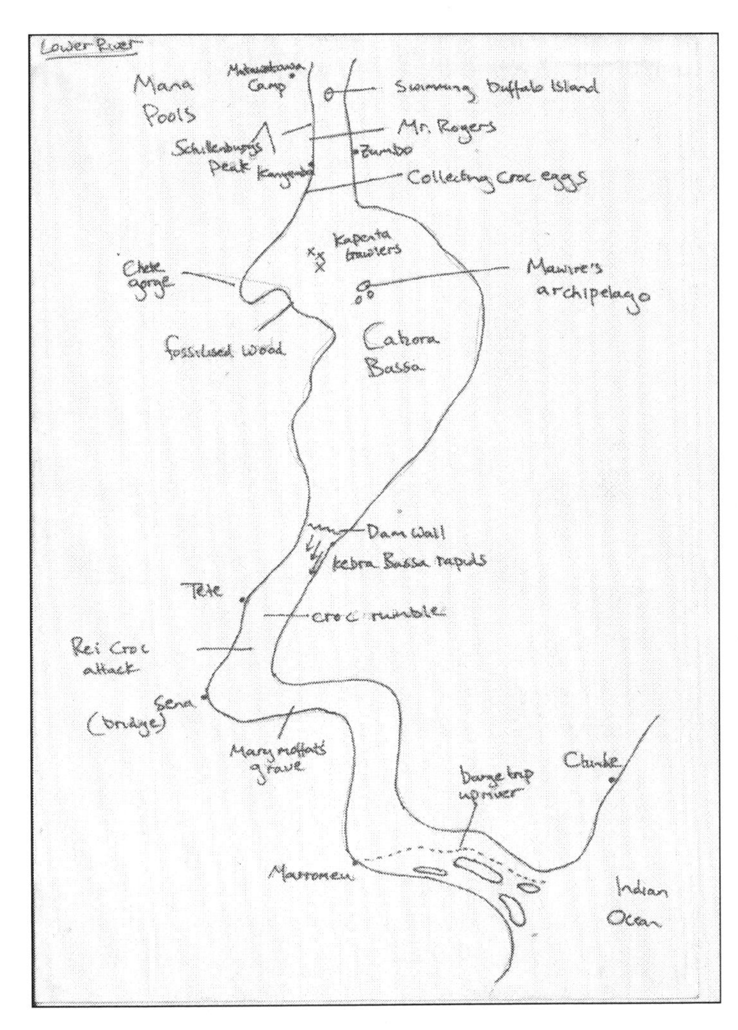

Sketch Map of the Lower Zambezi River

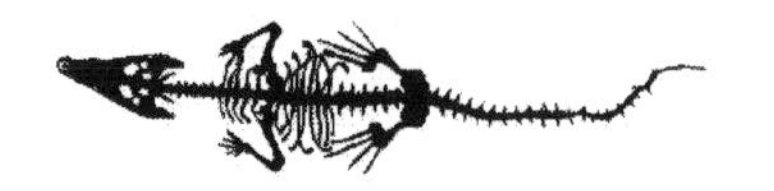

Chapter Fifteen

Mutawatawa Camp – Cahora Bassa Dam

Thou shalt steal only some of the eggs from a nest
– be it crocodile or bird

We began afresh, Reilly especially, after he hung his hammock between the skinning posts and had hyena sniffing around him all night! And once again our protective Tokoloshe was urging us on. 'One more river bend,' he said, 'and another one.'

We nosed out into the river and away from Mutawatawa into the beginnings of Mpata Gorge. There were a few dusty, dry river beds that came down into the U-shaped gorge. We could imagine that in the floods the gorge and river flow was so very different, even awesome. Where now, the river flowed clear and there were no sketchy water features, in high water the water would rage brown with sediment and there would be many lethal sieves and rock jams, features that our level of kayaking would be unable to deal with.

On the first night away we found a small island, almost smack bang in the middle of the river. Its surface held rich green grass cropped low and there were eight or nine *albidda* trees each ringed with dense bush below. *Surely here was a good place to camp*? We thought. Here there were no elephants to reach for the *albidda* pods and we would make a thorn branch fence to keep the hippos at bay. Dusk was an hour or so away and the sun sat low in the mulberry sky. There was time to

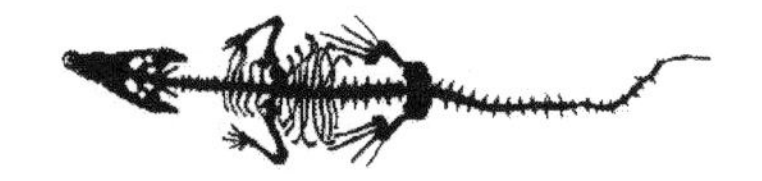

explore. It didn't take us long to find that we were presently sharing our island with five bachelor buffalo.

'They must have swum here,' Ace called from across the grass. 'Let's chase them off.' And so we began to round up the feisty males up as if they were mere cows.

Now, any old Africa hand will tell you, and rightly so, that buffalo are *the* most dangerous and unpredictable animals in the bush. 'Never,' they would say, '*never* chase or aggravate these thumping beasts unless you wish to play dice with your life.' I would never disagree with this but sometimes, and only very rarely, you come across an animal in such a state of mind that it has no desire to harm or kill.

Two or three years previously I had been on a solo trip up the Aberdare Mountains in Kenya. (By now you must have realised that this *really is* my most favourite place in the whole wide world!) I was in a homemade Suzuki that was known in the community as "the baked bean tin on wheels." Rounding a corner up on the bleak moorlands near the summit, I had come upon a lone bull elephant. He was about thirty and on the cusp of sexual maturity.

All around us were giant banks of mountain heather that grew in clumps around which the grass was chomped lawn like. Our impromptu meeting alarmed both of us, and with only metres between us, the ellie charged in reaction. I quickly judged that this was not a real charge as there was very little room and so I revved my little Suzuki and jolted

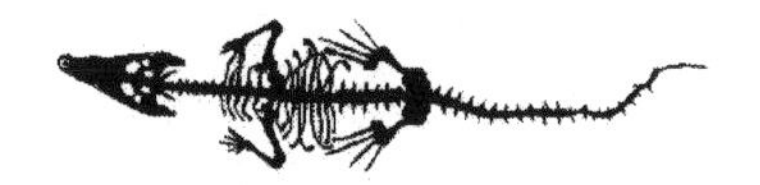

the 'tin-can' forwards at him. He halted and then reversed.

What ensued next may be considered rash, but in fact was one of the strangest games of hide and seek I have ever played. I reversed quickly around a clump of giant heather just off the road and *Tembo* came lumbering around from the other direction.

On seeing me, he pretended to have been frightened and would spin around trumpeting, trunk high in the air, ears wide and run back the way he had come from. Then suddenly there would be silence and I would creep round in my Suzuki the way he had gone.

The first time round, I went on instinct only. As I slowly motored round, I caught sight of *Tembo* tightly reversed into the heather holding the long branches over his head, obviously in hiding. I drove past him and pretended not to have noticed him. He was delighted, and just as soon as I had passed him he would come charging out making all the noise he could. He would mock charge the car and then caper off behind the heather again whereupon silence would once again ensue.

We played like this for thirty minutes or so until he decided that, energy spent, we were done. He headed off into the clumped heather and away from the road, and then turning suddenly, he made one more mock charge towards me and trumpeted as if to say *'thanks little yellow car that was fun ...'* and off he went.

These buffalo on the island seemed in a

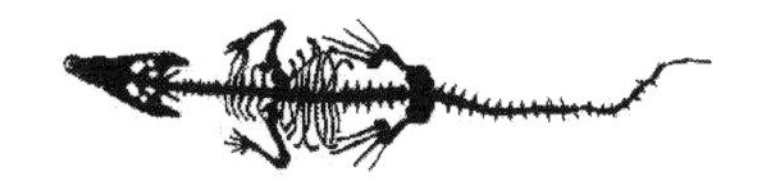

mood like the ellie, and so we gambolled about with them, chasing them from one side of the small island to the other. Sometimes they would bunch up and robustly gallop to one edge of the island where they would then break up and gallop about willy-nilly. They were frisky and were bucking and shaking their heads as little lambs do on the arrival of spring in the colder climes.

Finally they decided that they were done, just as *Tembo* at the Aberdares had, and gathering together they hurtled down to the beach and into the water and swam happily across to the mainland, five buffalo heads bobbing above the water. Once on the other side they turned and flicked their ears at us as if to say thanks for the game, and then they galloped off into the purple horizon.

By now the crocs had passed the breeding stage and the females were guarding eggs which meant that we had to be very careful which sand banks we pulled onto. Female crocodiles guard their eggs with valour uncommon in many reptiles; if their nest is threatened they can utter a literal 'roar' that is as terrifying as it sounds. The nesting crocs were just another reminder of how we had to read the river so very carefully, from hippos to crocs to water.

Years ago on the Mara River in Kenya, I think I was about nine or ten years old, we mistakenly caught a young crocodile!

Parents reclined in deck chairs as we children leapt in and out of a very muddy puddle that lay beside the river. We boys had discovered that two foot catfish lay buried in the mud at the base of the

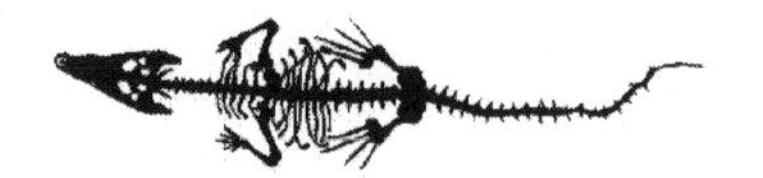

pool and were diving down to collect them, emerging from the filthy water triumphantly grasping these wriggling, whiskered fish in two hands.

'Mum look at me,' I called as I went down to catch yet another catfish. As I appeared from the water grasping my trophy the parents went wild and my catfish twisted in my grip and grabbed my finger in its razor toothed snout. I reacted quickly and hurled, what was not a catfish but a baby crocodile, quickly away!

My finger bled and our antics were abruptly curtailed by startled parents; where there were baby crocodiles there would surely be a mother!

Still we followed the Zambezi's flow and still we lay within Zimbabwe's borders. The Mpata gorge seemed wild and untouched with few human inhabitants. There were large granite rocks that jutted out of the water in pyramids; the walls were sometimes high, sometimes of shallow gradient. Here and there we saw the spoor of elephant, bushbuck, waterbuck, greater kudu, leopard and lion.

Sometimes at night we would hear a cacophony of baboon barking that indicated predators were around, and close. Everywhere there were raptors, and we were followed down the gorge by a pair of verreaux eagles for miles. *Ficus* trees clung to rocky shelves and hyrax frolicked in the jumbled mass of their roots, calling with alarm as we passed far beneath them. Behind these broken masses of rock the interlocking spurs headed backwards into the more wild regions of Zambia and

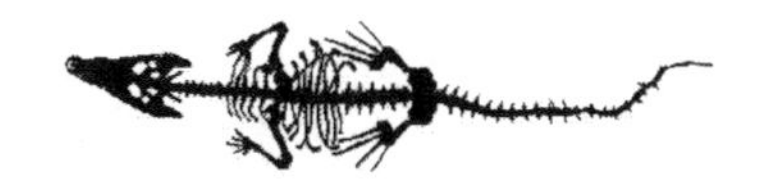

Zimbabwe.

Where thorn scrub could grow, it had taken a hold and claimed parts of the rocky cliffs for its own. Here and there, where the cliffs opened out in protected bays and between the spurs, large baobabs grew bearing the marks of many elephant tusks, perhaps even from thousands of years ago.

Along the riverbanks there were big, shady green trees doing well on the constant water. Electric catfish lay hidden in the pools of silt along the riverbanks and not surprisingly it was I who was our first and last expedition man to be stung by one when I tested the story by laying my hand across its body.

'Hey guys have you ever heard about electric catfish?'

'I have but I've never seen one' Rei called as he paddled over to me, 'you found one?'

'I think so, come see.'

Ace and Rei paddled close.

'Looks like an ordinary catfish to me, although what species I don't know.'

'Well touch it and see if it electrocutes you.' Rei and Ace beamed widely at me and then we all looked down into the water at the cornered critter.

'Yeah I don't think this one is electric,' Ace agreed as I leaned down and cupped my hand under the be-whiskered devil.

'Agggggghhhhhh! Oweeeee!' I whistled and hooted through my teeth, pins and needles shooting through me. I shook my hand violently up and down as the pair of them howled with laughter.

Large, rotund acacias, figs and jackal berries grew alongside the water and there were amazing campsites to choose from where we could have stayed a month and still not finished exploring all the hills around. The campsites lay between the spurs and were shady and cool with rich alluvial deposits from thousands of years ago. Nutrient-rich grasses grew and attracted grazers from miles around. We regularly saw snakes and spur fowl while the guinea fowl were always complaining about something. The hippos were an important part of the ecosystem because their dung was the fertiliser for the whole food-web that existed on each river bank.

Monkeys would be in the trees picking flowers and fruit, the eagles were in the sky waiting for the right moment to grab a monkey or a hyrax. To camp and hear the sounds of the owls and the nightjars, now the fiery-necked nightjar starting to call, it was thrilling. Indeed it was enthralling how much the night sounds had changed as we migrated down river: if you woke up in the night you could listen and almost pinpoint where in Africa you were, just by the night-time calls.

As we lazily paddled below the ramparts of the Muchinga Escarpment, flashes of deep pink glided between the mud banks – carmine bee-eaters. The Luangwa River spilled in from the North and we soon came upon Shillinburg's Peak, a literal pyramid with a miniscule triangular top. Reilly's

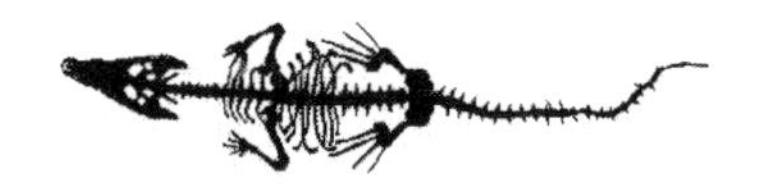

mum, Judy had told us all about this peak and Chris Shillinsburg who was decorated highly in Zimbabwe's bush war. Story held that this brave soldier operated alone on this peak. He dropped in by parachute to observe the 'terrs' (terrorists) and managed to get a radio message out to his commander of his observation. We had no idea of the battle that subsequently occurred, but in our imaginations we created graphic scenes and we spent that paddling hour and several hours later around the campfire that night, re-telling the story until we believed that we too could have performed the feat of a parachute landing on that pointed peak.

The lower rock faces of the peak were perfect for leopard and we kept a beady eye out in case we should be lucky to spot one. Scanning these rocky slopes reminded me of a leopard that had charged me in almost precisely this environment. Working in Makalali meant early mornings and late nights, but almost all the hours were filled with wildlife sightings that thrilled. Boorman was my partner in crime and in 'down-time' we were left to our own devices. May was low season at the lodge and few if any clients came through. In early April we had spotted a leopard with cubs and now we monitored her each week. One morning Boorman and I were sent out to find her location; there were guests on the way. We radioed in to camp.

'Trackers to lodge, do you read?'

'Come in trackers, we were about to call you anyway.'

'Leopard located, she's at Vlei Kopje, looks about to hunt over.'

'Thank you but we have just received notification that guests have cancelled. The day is yours trackers. Over and out.'

No guests and a leopard on the prowl; a perfect morning. We quietly drove the open topped Landcruiser deeper into the scrub at the base of the kopje, switched off and waited. Before us two male impala grazed sedately. We were deep into rutting season and before long the impala had locked horns and were grunting in the cool morning. We sipped from thermoses of hot, sweet tea and bided our time.

Suddenly there was a blur of golden rosettes as the leopardess streaked across the open ground from the cover of the bush. Before we even had time to exclaim the scene before us had radically changed. Gone was the calm morning, the singing birds, placidly rutting impala. Now the birds screeched alarm calls and as one impala fought for his life, the other desperately tried to unlock its horns.

Finally our leopardess won.

She dragged the male to the ground and grasped his windpipe in her teeth starving him of oxygen and bringing an end to the future survival of his genes. The leopardess was by now quite used to our presence and we watched her as she ate from her kill, not bothering to drag it away from us. She had made the kill in her home territory. After some hours, the sun now hot on our backs, she retreated, belly almost dragging on the smooth stone, moving upwards to the shade that the boulders offered.

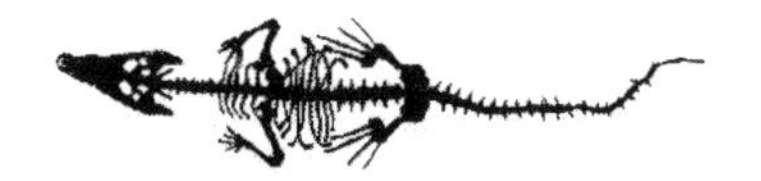

'Sergeant,' I said, 'how do you feel like a lunch of roasted impala?' To each other we had always been '*Sergeant*.'

'Seriously?' Sergeant looked at the kill, at me and at our leopard. 'OK here's my knife, I'll watch the leopard while you cut the meat.' Sergeant carefully repositioned the car putting it between the kill and the leopard.

I snuck out of the car and taking the hindquarter in my left hand I began to cut it away at its base.

'Have you still got your eye on her?' I whispered up to Sergeant.

'Uuuhh Sergeant I've just lost her, she's moved.' He paused to search frantically with the binoculars. 'But I don't think she's headed this way, I'd spot her.'

I had only a single tendon left to sever when suddenly, growling and spitting in a fearful and terrifying rage she tore from the bushes right at me – I leapt to my feet and keeping one eye on her I sprung backwards, as though I were Tigger, into the Landcruiser whence Sergeant floored the accelerator and we shot off – backwards and into a tree!

But the leopardess knew exactly what we had been up to and her charge had been a mock. She stared us down for a few seconds and then calmly wandered back to her sentry post on the smooth rock and gazed demurely down at us. The Cleopatra of the bush.

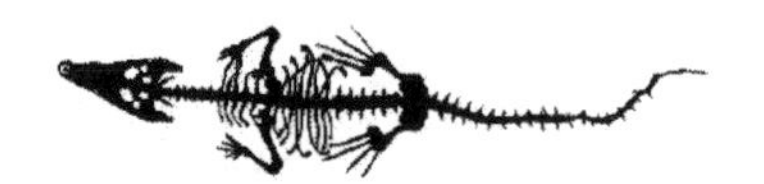

We spotted no Cleopatra on Schillenburg's Peak though we scanned it long and hard through binoculars. As we drifted past the smooth slopes of rock, piled here and there with mounds of boulders, large and small, the Zambezi seemed to be sluggish, weary of its travels and aware of the pressures it faced. It remembered the times of old, the hunter-gatherers that paced its banks and the battles that had been fought. But these were memories of yesteryear, long since carried away in the current to the ocean. The waters seemed to pull us along too, willing us to join the memories ahead of us.

There were a few extremely poor villages just before the Mozambique border that seemed to have stepped right out of the past. They had received no help from their government. The people were smiling farmers who irrigated their subsistence crops from the river. All night they would beat their drums to keep the wildlife from their farms; especially the crop-raiding hippos. Hearing the pulsing drums made us realise that the aggravation of the hippos in this region was due to this continual disturbance of their night feeding. To reach their grazing grassland they now had to by-pass many small holdings where they were hassled.

Life was basic here; one US dollar equalled a tin of worms which equalled fish for a week, while crops were jealously guarded as a failure of the short rains could easily end in starvation. Hippos were not to be tolerated.

We spent time here with the children who were keen to hear about rhinos and conservation.

The kids here knew a lot about wildlife and were open to ideas about how and why we had to protect. We made up songs with them that we all sang and danced to together. Reilly spoke KiShona to them and they understood. It was refreshing to finally have a positive response to our teachings of rhino and the importance of their conservation – but then this area had long since been poached out, the population still expanding beyond its means of self-support. Another rural community caught in the self-perpetuating cycle of poverty.

Kanyemba/Zumba was the border post into Mozambique and we made sure that we crossed at dusk, long after the officials had finished their shift. We had no problem crossing here and celebrated with a *mascotti*.

Our last leg had begun. Mozambique – Tete – The Indian Ocean! The excitement was palpable.

But first we had to cross the infamous Cahora Bassa, a man-made dam that is often called a lake due to its sheer size. Everywhere there were signs of hippo poaching. We spent that night in a maize field with near on the worst mosquitoes I've ever experienced. By this stage, our mosquito nets had holes in them and we had no hope in keeping these malarial blood-suckers at bay. We had to sleep under our blankets in a stifling, sticky heat.

Morning found us paddling alongside a high bank on river right, not too far into the dam, an obvious relic of the high river-banks that had been blasted away to make way for the rising dam waters. Habitation had thinned out and thorn scrub had crept back in. The water was brown and loaded with

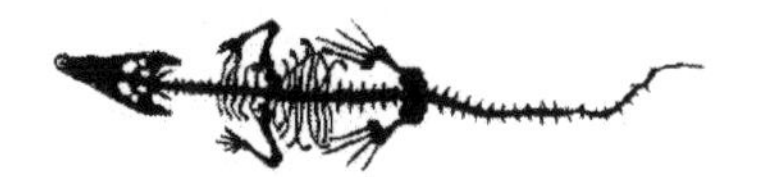

sediment and only a slight breeze whispered among us. As the banks fled away and the feel of being on a dam started, we met some *mzungus* (white men) who were line fishing. Hidden behind them was a fishing lodge with no guests. They invited us in and we were treated to biltong and stories.

'You know there were a hundred and twenty people eaten by crocs in Cahora Bassa last year (2008),' they said, 'so you boys better be careful hey. Don't do anything silly. Keep your wits about you.'

'I oath I'll do just that and more,' Reilly said firmly.

Come late afternoon we declined offers to stay and left the lodge in trepidation, our heads full of the stories of man-eating crocodiles. We crossed over and started paddling along the southern bank of the dam. The northern bank on the map looked a bit dodgy with long stretches of open water crossings. We had read that you had to take into account the wind which could blow you a long way off course.

An hour's paddling along the southern edge brought us to another camp where we met a very enigmatic and friendly Mr. Rogers. He insisted that we should stay.

'You boys need to see the landscape around, not just the water. *Ach* man, looking at that water all the time, it'll make you *penga-penga*.'

Upon the absolute insistence of Mr. Rogers, we spent four days of fantastic exploring up into the

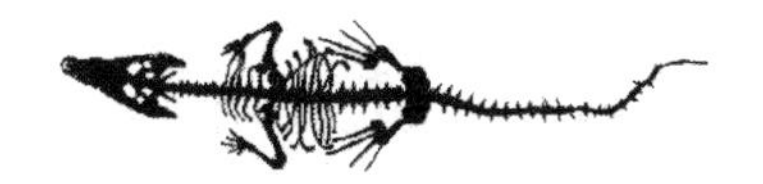

hills behind and amongst the floodplains. There were enormous *Brachistegia spiciformis* that sported stag head orchids, like fascinators. There was almost complete canopy cover. It was incredible; there were sable, eland, warthog, buffalo, giraffe, zebra, lion, leopard and wild dog.

Many species had been severely reduced in number as a result of Mozambique's civil war and poaching had also taken its toll, especially on sable. Sable are incredibly beautiful animals – the chestnut hide of the females glows with health and darkens as they mature while the males are a sleek black that contrasts almost poetically with their clean white underbellies, cheeks and chins. A good male specimen with have a shaggy mane and both males and females have beautifully ringed horns that arch gracefully backwards, for up to a metre. Sable fight their predators using their scimitar shaped horns, but these serve as no protection against trophy hunters who desire only to mount the heads of these magnificent antelopes on their walls. These numbers were also drastically reduced when hundreds were culled to try to eradicate the biting tsetse fly.

Mr. Rogers had taken a ninety nine year concession of an enormous tract of land and it warmed our hearts to see that he was nursing the land back to its natural health and wealth. His son too held the spirit of Africa and it was he who would take the conservancy over.

One evening as we wended our way back to camp in the Landcruiser, we came upon a young male leopard that was scent-marking his territory. His golden spotted pelt glowed in the headlights as,

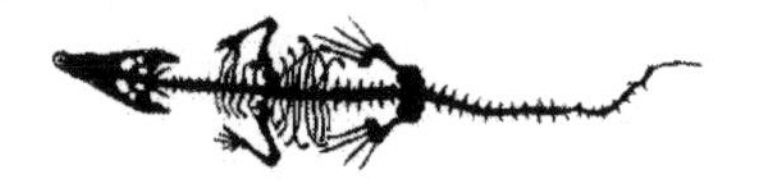

unhurriedly and without flinching he swayed in front of the car, moving from side to side over the dirt track, marking first one bush and then another in his claim to territory. After ten minutes he disappeared into the shadows as suddenly as he had appeared.

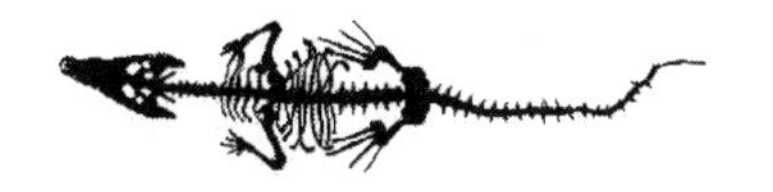

Chapter Sixteen

Cahora Bassa Dam – Kebra Bassa Rapids

Thou shalt never take the river in vain

Sketch Map of Cahora Bassa Dam

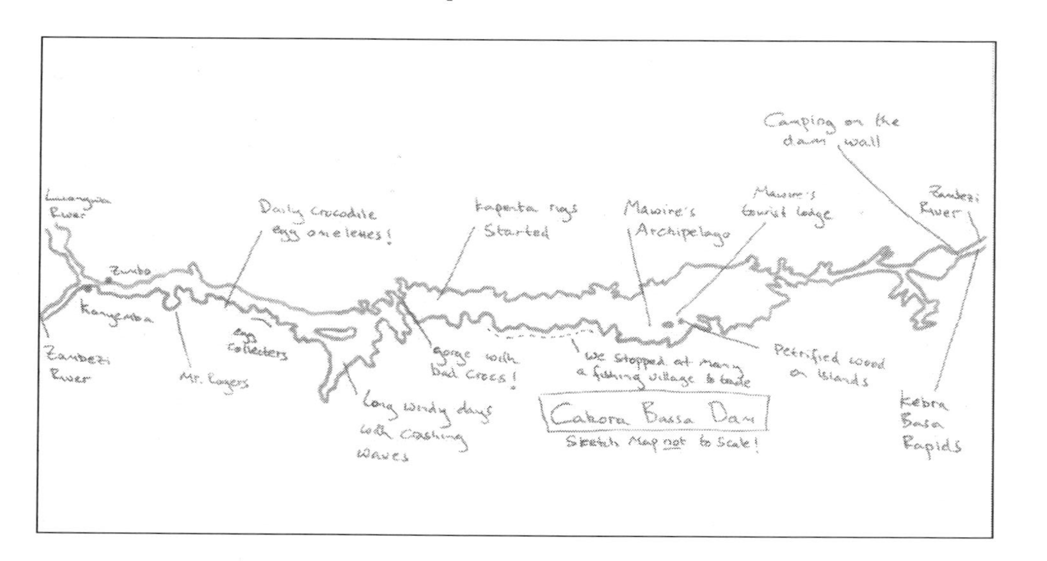

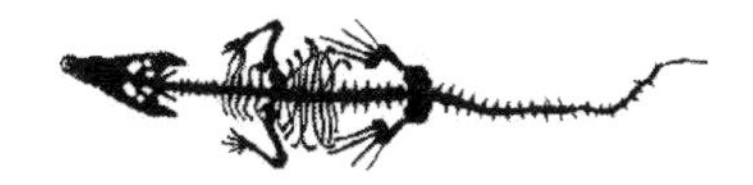

Before independence, Portugal fought bitterly with the Frelimo independence movement; then in 1975 Mozambique got its independence from Portugal and was almost immediately plunged into civil war, famine and widespread corruption and mismanagement, of land, resources and people. After sixteen years of civil war, Frelimo pitted against the Rhodesian and South African sponsored Renamo movement, a peace deal was signed in 1992 and slowly economic development and political stability crept back in. But the country was in chaos; over a million people had died in the civil war and millions more had fled across the borders.

In 1990 Frelimo inaugurated a new constitution that enshrined free elections, and although Frelimo has won all subsequent elections, political life has remained relatively stable. Renamo continues to work within the constitutional system and the country has a growing economy. Although there are natural resources - coal, titanium, oil and gas - the majority of the population works the land. War, neglect and under-investment have left a country bereft of infrastructure, and floods and droughts mean that poverty remains widespread with more than 50% of Mozambicans living on less than a dollar a day.

We had been told that Mozambique had been 'poached out;' that the war had decimated wildlife populations throughout the country. Already though, we had been lucky to see a fair amount of game. As we carried on along the southern bank of the dam, we had to avoid these enormous sand spits. There were very few islands and we had to do more long days because of it. We would stick about a kilometre out from shore and paddle on a bearing for twenty

kilometres or so before we would come inland. If we had not done this, we would have zigzagged along the bays which would have doubled our mileage across the dam.

On shore we would stop off and gather handfuls of crocodile eggs with which to make omelettes. Boiled croc eggs turned out to be pretty rough; a leather texture that rasped against the tongue. Scrambled crocodile eggs, on the other hand, were great with our morning tea, and would give us the boost of energy that we needed for the day's paddling.

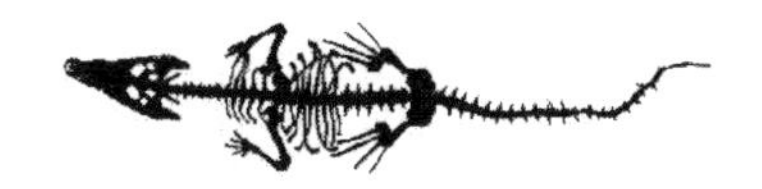

Crocodile Egg Omelette

Omelette is a dish that can be served anywhere at any time of day. It is easy, nutritious, and versatile. It reminds one of being at home ... oh for a non-stick pan!

Eggs are delightful when appetite for all else has faded, and with chilli and fried fish pieces an omelette will go down very well indeed.

Omelettes work as a treat for breakfast and for elevenses too if you are hungry. You can always add chunks of venison and slices of potato. Of course cheese is always an added bonus but out on the river it is often not available.

Serves 3 hungry but lazy kayakers

8 x crocodile eggs
1 x potato
Venison, if available
Salt, pepper and chilli

1. Collect eight crocodile eggs.
2. Beat the eggs with a fork and season with salt, pepper and chilli.
3. If you do have potato or venison to add, then slice them thin and fry the slices lightly first.
4. Pan fry on medium to hot coals.

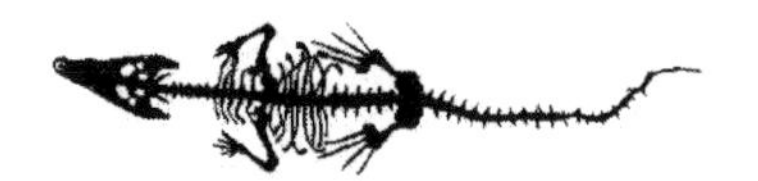

On the sandy banks of the dam we met some egg collectors. They had opened a new breeding facility and were building their base population within the farm. They intended to collect around fifty thousand croc eggs that they would incubate and hatch at the farm. After this initial collection, 80% of breeding would then be done 'on site'. These farms bred crocodiles for the skin market in Europe (mostly France), and for the meat market in Europe and Asia.

Local laws require that a certain percentage of crocodile hatchlings are released back into the lake ecosystem to sustain the wild crocodile populations. This policy is controversial as local communities argue that crocodiles threaten both livestock and humans.

The community here had threatened to burn the breeding centre if they released any crocs back to the lake and I suppose it was the same in many other places as well. We didn't ask if they collected fewer eggs as a counter-method.

It took us a full week to complete the top reaches of Cahora before we started coming across the kapenta fishermen.

'Look out for Mawire' we had been told. 'When the kapenta rigs start, he could be anywhere.'

'How will we know it is him?'

'Oh you'll know; Mawire is not exactly your norm. He's expecting you anyway.'

On the very first rig we met an employee of

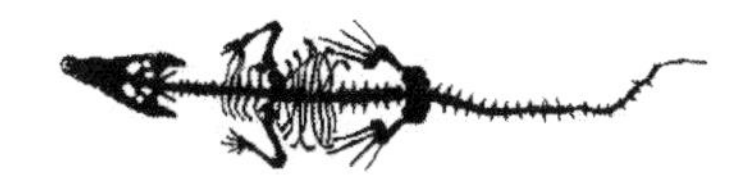

his who pointed us in the general direction of Mawire's archipelago of islands.

But the kapenta rigs were interesting and warranted further investigation. Although there had also been kapenta fishing on Kariba Dam we had not taken the opportunity to find out how the kapenta rig components are made, how they are fitted together and how they work.

Kapenta represents two species of small fish, the Lake Tanganyika sardine and the Lake Tanganyika sprat. Both freshwater species are planktivorous and pelagic and form the major biomass in the open lakes of Tanganyika, Victoria, Kariba and Cahora Bassa where they were introduced. In many regions these species have been over fished and now even more businesses have sprung up around Lake Victoria in East Africa, exporting these small fish south.

The fish are caught using specially fitted rigs which use mercury lights powered by portable generators. These attract the fish above a dip net of a six metre diameter. The nets are then drawn from a depth of about forty metres.

Kapenta breed in the shallow waters and in an attempt to halt overfishing many countries have made it illegal to fish in water shallower than twenty metres.

Kapenta are eaten after first being dried in the sun on racks and then typically served deep fried. Sometimes the fish are salted but this adds to the end cost and as it is the staple diet of the poor, this is not preferred. Kapenta is a refrigeration-free protein that can feed an entire family along with

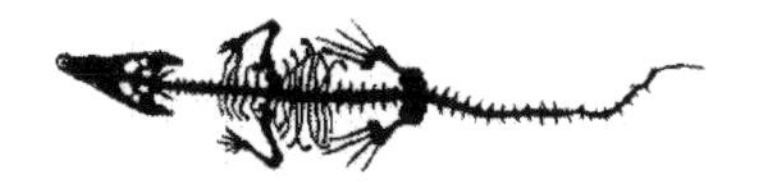

onions, tomatoes and *sadza*, the maize meal carbohydrate that is popular in southern Africa. In better-off house-holds, the kapenta can be marinated in vinegar with salt and kept in a refrigerator. After two or three days the vinegar is discarded and the fish is quickly rinsed with clean water. The 'fillets' are then put in a mix of olive oil, vinegar, sugar, garlic, chilli and lots of parsley or celery. After another two or three days in the fridge the marinated 'fillets' are ready to eat.

Having got to grips with kapenta and having eaten enough to last a while, we pushed on across the choppy waters of Cahora. To get to Mawire's archipelago required that we pass a south-shore gorge that we had been warned about. This gorge was renowned for the attacks of a six metre croc! Our paddle strokes were clean and fast as our eyes continually searched the water line but, (thanks to our *Tokoloshe),* we didn't meet him.

As we paddled this section the wind buffeted us from all angles. The waves broke continuously over our kayaks and made us nervous. We did not want to get stuck out here. Each situation could easily turn life-threatening. Our safest game plan was to wait for paddling windows when the wind had dulled down, but these followed no particular pattern and it took us several days of hard paddling to reach Mawire's islands.

Several times over the days we pulled in at fishing villages where we would swap fish hooks for staples. Sometimes we stopped off at islands to explore. The greatest bush-telegraph of all preceded us - any time that we came upon kapenta boats we were recognised!

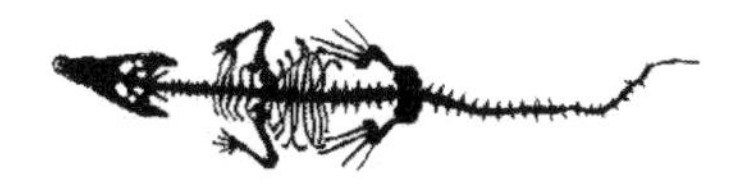

Mawire had told the fishermen to watch for three kayakers and had instructed that our bellies be filled and our thirst be quenched. And so it came to pass that breakfast, lunch and dinner were kapenta!

Finally we came within sight of an archipelago that, by description belonged to Mawire. A low throbbing noise came from the air and as we paddled, eyes to the sky, he began to dive-bomb us in his micro-lite. Wild shoulder length hair that had never crossed paths with comb, lent character to his strong frame of forty odd years and beach attire lent the colour. There was certainly no problem in recognition!

Mawire turned out to be a child of the wilds just like us. The only difference was that he played this game all year round. Johan Hougaard no longer suited as a name, the village people had coined Mawire and so it stuck.

Fair in his dealing but short in fuse, he was easily upset. Many a tale had been told of Mawire and his islands on the great Cahora and the longer we spent with this eclectic man the more real and animated these tales became. He lived life at high speed, flew a lot and did nothing without full zest.

Mawire was glad of the company and all night we told stories, each more wild and exaggerated than the last! Tales of outrunning crocodiles, wrestling with pythons, crabbing amongst hippos and boat crashes! Yes there was many a tale to tell of bruises that still throbbed!

More than anything, Mawire had wanted a

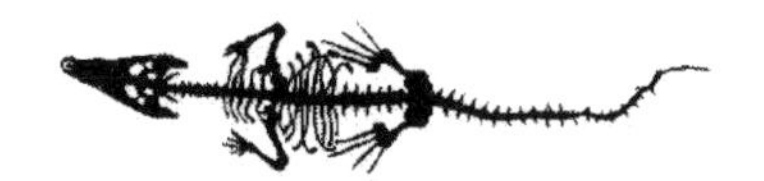

son to raise in this life that he loved so well, but sadly it was not to be. After several marriages and still no sign of a '*lightie*' on the horizon he took to life on his own. He lived on this small island and for earnings, ran kapenta fishing rigs. Fear was not a word that was in Mawire's vocabulary. He pioneered everything and anything – built and designed and tweaked. He woke early each morning and slept late each night, living each day to the full.

Mawire took us flying in his microlite, soaring over vast areas of the lake at dusk with spectacular views across the twinkling lights of the rigs. The microlite is the motorbike of the sky – you feel wild and free! Rei hinted at captaincy, but there was no way we would let him take the helm, no ear-ripping for us thank you!

Island life held us captive for a good week or so. We watched and learnt as Mawire showed us the inner working of the kapenta business; we soaked in the warm waters of the lake, fished from rocky outcrops, drank rum and coke under the setting sun and dozed in our hammocks as the breeze serenaded the aroma of drying kapenta. As the days crept into the middle of October we realised that both Mawire and I had upcoming birthdays.

An impromptu party happened on the Saturday after a fiery Friday night sundowners. The guests were few but the entertainment was merry. We were joined by Mawire's brother who showed that amicable madness ran strong in the Hougaard family. The siblings worked together - Mawire's brother organising the kapenta fish sales from Harare. With a strong interest in geology and

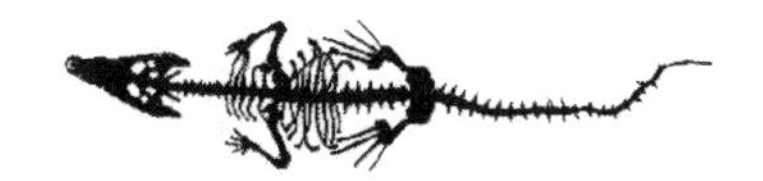

fossilised woods it was he who told us of the fossilised forests of Cahora Bassa. He was an audacious partier whose girlfriend at home was younger than his son from his previous marriage!

Mawire's house was designed to entertain. Funky wooden pieces, collected as flotsam from the lake had been polished up to make tables and bar-stools. The bar itself was made from all of Mawire's crashed microlites and his broken boats. Everywhere there were fancy hats hanging and together with the sound system and karaoke machine he had installed, the party grew pretty wild. We danced and we sang, we drank whiskey and we told stories, and then we sang and danced again. There was no screen to follow karaoke words on so we made up our own songs, while Mawire's brother played the guitar. We sang about "Row Rhino Row" and songs about the Kebra Bassa rapids that we were about to do.

Hung-over day followed hung-over day until the wind whipped us into action. Between our merry evenings we felt we *must* explore!

Following directions from Mawire's brother we found a large section of fossilised forest in a collapsed bank. Being the collectors that we were, we all wanted to take a piece home and no-one was going to settle for a small one. It had to be a whole root or branch. We collected them and paddled back to Mawire's – exhausted but happy after a days excursion and ready for another night of wild karaoke!

The water around the island was clear and spear fishing easy but risky. Two would stay on crocodile watch, one with the boat ready. We had

bangers we could let off underwater that we figured would scare any croc into dropping its victim. We roasted the fish we had caught in an oven with spices and olive oil and salt, growing fat on the spoils that we caught.

But time stops for no man and soon we knew we had to leave this Eden and its host. Ahead were the rapids below the dam wall and Mawire wanted in on that adventure!

We left the island having agreed that Mawire would meet us at the dam wall. We would then drive together around the gorge below the wall, to a small camp that Mawire kept below. From there we would tackle the rapids in the upward direction in the speed boats and at the top would jump into our kayaks and paddle down.

It took three days to paddle to the dam wall and we spent our nights on other lake islands, one being where Mawire had built a tourist lodge. The island rose steeply from the water and had been badly eroded. These 'islands' would have been hilltops around the river before it was dammed. There were amazing formations of different rocks below which the lake's bottom fell off sharply into deep, clear waters.

We eventually reached the dam wall through a myriad of dead ends and bizarrely shaped waterways. It was quite a mission. Behind the dam wall the hills rolled like ocean waves, far and away, down towards the Indian Ocean.

Portaging our kayaks up the steep slopes to the dam wall, we found Mawire waiting for us.

As dusk came we looked down upon the twinkling lights that ran the length of the dam wall. Black silhouettes of rocky crags and hills showed themselves to still be wild as hyena and leopard calls echoed between them. There were signs of recent movement of elephants and people seemed to be few.

'Plenty of unexploded land mines here,' Mawire informed us. 'Got to be *bloody* careful where you walk.'

Finally we lay exhausted and sun-burnt in our hammocks. The hard slog of Cahora was over. My eyes closed and my tired muscles rested as somewhere, far below us a deep 'boom' exploded into the night.

As the sun rose the following morning we loaded our kayaks into Mawire's car and dropped down into the valley.

'Got to leave before it gets too *bloody* hot,' Mawire hurried us to finish loading as he slapped the tsetses away. We grinned as he put away few slugs of Famous Grouse. 'Medicinal.' He stated. 'Keeps the tsetses off your blood - or at least, keep them from finding the next place to bite!'

We seemed to drive through solid walls of heat. No cooling breeze evaporated the beads of sweat as they ran in rivulets down to our mouths.

Salt on the tongue, we stopped for lunch. Mawire set up a bench table in the tepid water and we submerged our bodies from the biting tsetses and swigged cold beers over cold meat sandwiches

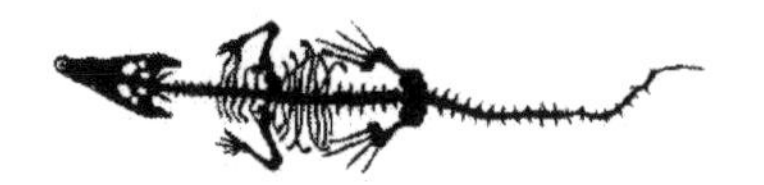

and biltong.

Around us jagged rocks, moulded by time and water, towered up into the hillside. The land was harsh and not people friendly. We pushed onwards, our kayaks in the boat and the boat on the trailer as from the water we climbed back up the valley sides on the roughly hewn track.

In Mpeza Nyota, a small town soon after lunch, Mawire met a great old *rafiki* of his who had previously robbed him of a whole Kapenta trawler and sold it. All that was now behind them both, and this old pirate was a boozer who was full of fun. Here too we picked up Mawire's nephew who was keen for adventure and would be a great helping hand in safety for the rapids to come.

From Mpeza Nyota we turned East down an old severe, seldom-used, wash-away road. The road did nothing for the car and we had to change punctured tyres and do repair work on the trailer tow-hitch; the nuts and bolts had sheared right through with the strain being placed on them. Eventually we could, once again, see the river between huge baobabs and sand gullies.

Entry in and out of the water access point was tricky and our wheels fought hard to gain traction. Here the river ran sedately past denuded land cleared of trees for its penny charcoal value. Soil erosion created deep gullies that fed silty rain wash into the muddy river that shared none of the tales of the rapids we were to face upstream.

Wading in the water, watching carefully for crocodiles, we offloaded the speed boat and our

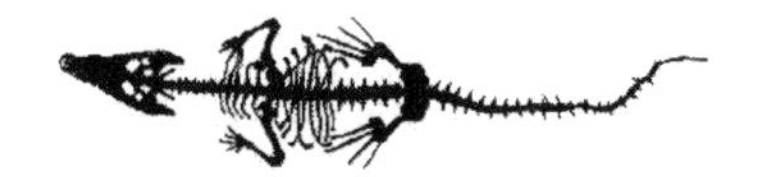

kayaks. Wheels spun and we added to the erosion as the car pulled away from the river. It would have been nice for the *askari* to have been left to guard the car in shade but not a tree remained to leave the car under.

Mawire had a small 'camp' on the other side of the river and we sped over in the speed boat leaving car and guard to scorch in the sun until we returned. The camp was very basic; tents, bucket shower, long drop toilet. The caretaker met us with a wild grin, obviously pleased to have company.

'Eh! You have come, it has been sooo long!' he whooped and shook all our hands solemnly before setting to work lighting a fire for tea water.

Francolin calls rent the dusk air and Mopane trees shook their butterfly leaves at us as we rigged our hammocks under their shade. The firelight danced to and fro in the dark, sometimes bringing Ace's face into focus, sometimes Rei. We needed to be well rested tonight if we were to pull off our audacious plan tomorrow and it wasn't long till we trailed to our hammocks water bottles in hand.

My mind conjured up hidden rocks in the river gorge and they caught and damaged my kayak, flinging me out to make my own way down the thundering white water. I pulled out of my reverie and reasoned with myself: Mawire reckoned that he could remember how to reach the dam wall safely. Mawire and his nephew would keep close by us should we need rescuing.

The descent of these rapids could easily kill us but Mawire's bravado had spurred us on.

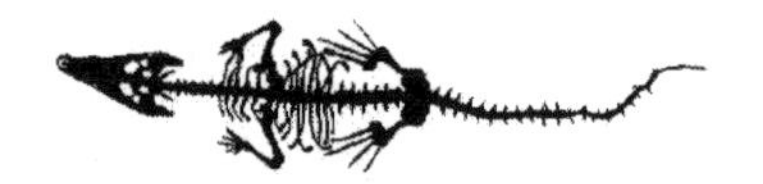

Everything was possible.

The next morning we awoke to the sweet smell of mopane smoke. It drifted aimlessly around us in sloping question marks as we chowed down breakfast.

It was time to set off up-river. My heart hammered. I looked at Ace and Rei; I could see they too were nervous.

We loaded one kayak in the small speed boat with me and Mawire's nephew at the helm. The other two kayaks were towed behind Mawire's big 150 Yamaha engine boat that held him, Ace and Reilly.

We checked the boats over one last time

'You follow me, ja?' Mawire instructed as I revved the motor of the smaller boat.

There was no option but to take the upstream journey at speed, only eddying out when the waters allowed us to. The waves threatened to swallow us and the boats screamed their watery anguish.

Early on we encountered foaming waves that engulfed Mawire's boat fully as he throttled hard against their power. We eddied out below and waited for our turn. Our boat was smaller that Mawire's and I wasn't sure we would make this run.

'Get some speed up,' Mawire's nephew cautioned.

I pulled out into the flow and let it take me

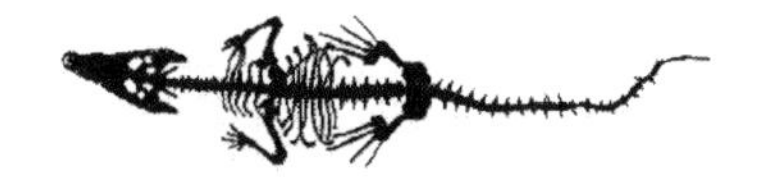

downstream before I turned back into the flow, opened the throttle and went full speed into the foam. The engine groaned and went silent. The sudden loss of power allowed the water to turn us sideways and we fought to face the waves again.

The engine spluttered and above in the calm water I could see Mawire, Rei and Ace watching intently. The water again broke over us and violently twisted us about ripping the engine from its mount. It fell into the water and disappeared!

We were now at the mercy of the rapid and with nothing to keep me pitted against the angry water, we were now swept downstream with no control. As we were dragged with the flow towards river left the boat slammed into a rock and water began to flood our feet. There was a hole in the hull.

Mawire's nephew bailed frantically but the water rushed in, threatening and trespassing indiscriminately. Somehow in our panic we had not seen that Mawire was back beside us. A coiled rope landed on our deck and by instinct we both snatched for it.

The boys dragged us into an eddy where the boat went down. It was an event we had semi-prepared for. We had tied large ropes with *mitungi*s (jerry cans) to the boat sides, so that if one or other of the boats did go down we would have markers as to where. We used this rope now to drag the boat more towards the river's edge.

Mawire insisted the adventure continue.

'Rope her to a rock; we'll come back this

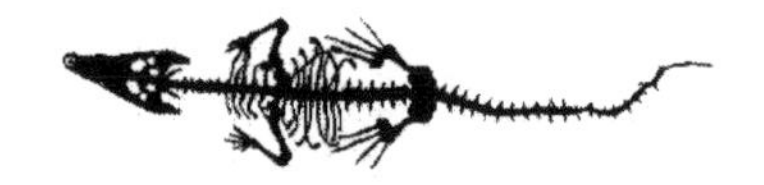

evening.'

The kayak we moved to Mawire's boat and somehow we managed to make space for the five of us.

Again we pitted ourselves against the Kebra Bassa rapids. The two kayaks that we were towing did not stay upright and at various times we had to stop, empty them of water and right them. Sand ballasts kept them somewhat steady.

Everywhere there were rocks that jutted out into the current, rocks that we had to avoid at all costs. Mawire had by this stage, drunk his morning half bottle of Famous Grouse from the twelve bottle box that he had brought on the trip. As we made good progress on a somewhat flat section of water, the boat suddenly careened to starboard and the bottle of Famous Grouse was almost washed over.

The power steering had broken.

'Not *lekker* bru!' Mawire muttered under his breath 'Ace grab the outboard we're heading for shore.'

Ace held tight to the outboard in case it was ripped away from us by the powerful flow. As we reached the shore, Rei and I leapt out and pulled us in.

'Nothing we can't fix,' I yelled too loudly, my ears still ringing with the roar of the water.

It took us thirty minutes to fix a shaft onto the engine.

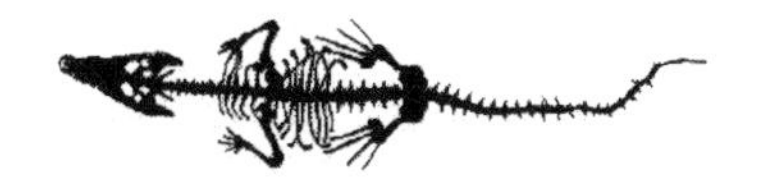

'Right Ace, I'll stand at the helm and work the throttle,' Mawire called, 'I'll shout directions to you and you work the engine like an outboard.' Ace grinned and we climbed aboard for the final up-river push.

Somehow fate looked kindly upon us and we made the dam wall without further incident. We found some still water and a small sand beach to pull up on. We needed some lunch, some beer hydration and to mentally prepare for the trip back downstream, although already we were ruminating on the fact that we were still alive after the upwards stretch!

In the forty minutes that we spent below the wall, we had enough time to cast a few hand lines that Mawire had wisely stowed away in his boat. Word was that the fishing below the dam wall was legendary. Within five minutes, first Ace then Reilly then I hooked an eight kg tigerfish from the oxygenated, deep clear water.

We had reeled in twenty-four kilograms of tiger in three casts!

Mawire stowed the fish in the boat and we suited up for the downstream kayak adventure - one we had to complete without Eskimo rolling as we still had not learnt that trick!

Livingstone tells of the rapids in his book, *Expedition to the Zambesi and its Tributaries*, *'From what we have seen of the KebraBassa rocks and rapids, it appears too evident that they must always form a barrier to navigation at the ordinary low water of the river; but the rise of the water in this*

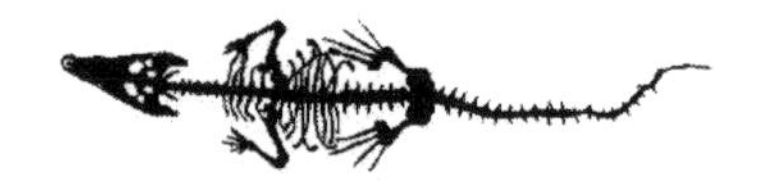

gorge being as much as eighty feet perpendicularly, it is probable that a steamer might be taken up at high flood, when all the rapids are smoothed over, to run on the Upper Zambesi. The most formidable cataract in it, Morumbwa, has only about twenty feet of fall, in a distance of thirty yards, and it must entirely disappear when the water stands eighty feet higher.'

Livingstone goes on to write that *'it is currently reported, and commonly believed, that once upon a time a Portuguese named Jose Pedra,— by the natives called Nyamatimbira,— chief, or capitao mor, of Zumbo, a man of large enterprise and small humanity, being anxious to ascertain if KebraBassa could be navigated, made two slaves fast to a canoe, and launched it from Chicova into KebraBassa, in order to see if it would come out at the other end. As neither slaves nor canoe ever appeared again, his Excellency concluded that KebraBassa was unnavigable.'*

Now we found ourselves in our more modern kayaks and filled the different compartments of our empty kayaks with sand to balance and add weight. With life jackets and spray decks, each of us got comfortable.

Was I worried you ask? I was terrified ... *What were we doing?*

Mawire and his nephew were to follow close behind and to rescue us if we needed. With trepidation, we ferried out into the current.

My diary recounts some of the run down: *'At*

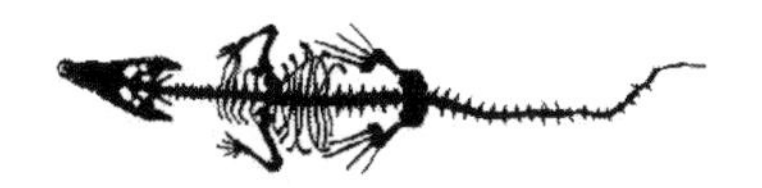

various times we would come up to these whirlpools and they would just suck us in – I remember frantically fighting these things and generally you were just not winning, rather you were making yourself more and more tired. By the third or fourth whirlpool your only aim was to keep your nose out. Luckily none of us was dragged further down than we could fight. Some of these whirlpools came out of nowhere and I am sure could suck us in and we would be gone!

Ace did roll at one stage and had to pull his deck. He first managed to by-pass a mean whirlpool that sucked his boat down for at least fifteen seconds before it was spat out. I was on the same section right behind Ace and tried to loop a rope onto his kayak so I could grab the sucker and pull it out, but I missed. Once the kayak came out it was up to us team mates to grab the kayak and pull it to the side, into a safe eddy, empty all the sand and water, put new sand in, get the weight right and begin again.'

A little after Ace's frightening incident I lost my nerve on a relatively simple stretch of rapids and capsized. The water took me down into a small pool below which lay a churning mass of jumbled white-caps. Paddling ahead of the other two I was headed straight for them. Alone.

'Jimbo grab the rope, I'm throwing it in three – two – one,' Mawire's nephew shouted.

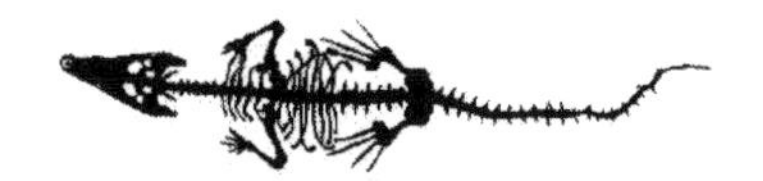

I gulped a mouthful of water as I fought to look up at him. Mawire had turned the nose of the boat upstream as I grabbed the rope, and like a jet-skier tows a surfer ahead of a wave; they pulled me out just as my feet were being dragged down the next set of rapids. Coughing and spluttering they hauled me aboard and still facing upstream we navigated our way to the much larger pool section below where Ace and Rei were waiting.

Always Mawire was following behind, filming various escapades while still keeping the boat on line.

Wherever we could, we would peel into an eddy and have a break. In one eddy we came upon five or six locals collecting water. Whether they were startled by our madness or the kayaks and roaring boat, we don't know, but they dropped their *mitungis* and ran for their lives, zigzagging up and away over the high banks.

From another eddy we climbed up the gorge walls and Mawire showed us where he believed Livingstone had camped before they realised that the Kebra Bassa rapids were un-navigable.

In 1853, Kirk, Livingstone and Rae holed the *Ma-Robert* steamer just above the water line on their first attempt of the Kebra Bassa rapids. They then followed the river on foot and to Livingstone's dismay found a thirty foot high waterfall. (This does not now exist as in its place stands the concrete dam wall). The waterfall marked the end of a dream – a dream that one could navigate up the Zambezi for trade.

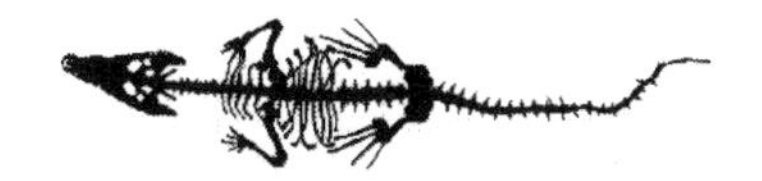

The ruins that we walked through were brick shelters. There were dry stone walls and hardwoods, bricks with plastered remains that formed broken walls. Hills protected the 'settlement' from attack from behind while the river formed a natural boundary in front of them.

It was a relief to be in camp that night, re-living and embellishing on our (luckily) successful descent. That night when I counted the stars to bring sleep to my still adrenalin charged body, I knew that somewhere up there were three lucky stars, one for each of us. Mawire didn't need a star; he had his lucky Famous Grouse!

Dawn found us still lazily swinging in our hammocks; today would be more sedate. After a tiger of a breakfast we loaded into the speed boat and pushed up the first rapid and onto the bank where we had to set to work on the sunken speed boat. We managed to pull the boat, minus its now drowned engine, out into the flow and floated it, semi-submerged down the rapids, dragging it to shore again at camp. While Rei and Ace searched the eddy for the outboard, Mawire, his nephew and I drained the water and pulled the boat up safe. By sheer chance the boys did indeed find the outboard and within six hours we had it running smoothly again.

We covered the boat and stowed it away high up the bank – ready for someone else's adventure. Mawire had to go back to work and we had to push on. In the morning we packed up camp and helped Mawire and his nephew load up the vehicles. We waved our goodbyes and as red dust

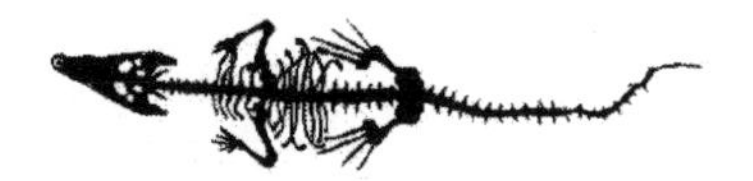

filled the heat laden air we began to paddle again.

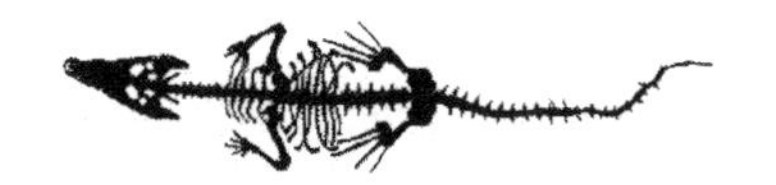

Chapter Seventeen

Kebra Bassa – Indian Ocean - Marromeu

Thou shalt repay all of nature's gifts with respect and protection

Below the gorge the river begins to calm again and the paddling was sedate and easy. The steep sides of the gorge fell away and several drier river beds met the water. We drew very close to Tete which means 'reed' in the local tongue. It is the location of one of only three bridges across the river in the entire country. Before Tete though, we called in at the Boroma Mission which was built by the Portuguese at the end of the 19th Century. The buildings are now used as a school and as the kids were at work, we were not able to explore all that we wanted to. The chapel interior is still preserved and the original fresco paintings still visible, but most everything was in a state of disrepair; holes in the ceiling, chicken coops in some rooms and monitor lizards in the rubble. Even a solid iron cannon had been broken in half!

Twenty kilometres further on, we came to Tete and its bridge. Tete had been an important Swahili trading centre in the past, and a centre for gold and ivory trading in the 17th century. The one kilometre long suspension bridge provides a vital link with Malawi and Zimbabwe as well as between the north and south of the country. As we passed by, the local kids would leap off the banks and swim out to our kayaks, grinning and singing. This worried us

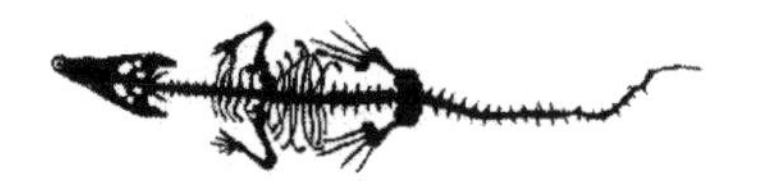

slightly as we knew there had been people, and especially children, taken by crocs right here, but the kids were unconcerned and gambolled happily about us. We stopped to re-supply and to have a cold beer and then decided that we would stay in a small *hotelli* overnight. Mawire had promised to drop my passport and fully legitimate visa off with a contact of his in Tete, and quite surprisingly we found the man and he happily gave me my paperwork over a cold soda.

The next morning we re-packed and continued on our way between the ramparts of the new railway bridge that they were building below the town. We explored several old ruins of indistinguishable origin and an Old Portuguese fort whose hieroglyphics, the man told us, had never, to this day, been deciphered. Many of the ruins had been looted of any metal that could be sold into the scrap metal market. In some places huge fig trees had grown into the stone walls and their roots dripped down the edges, the structure now crumbling from nature's force. Long, narrow islands split the river into two slow channels and in places the water was shallow and the white sand showed through the jade depths.

It was on this section that we began to be 'bumped' by crocs. We remembered Norman Travers warning us about breeding males mistaking the kayak shadow as a competitor, and so we made masts for our kayaks with our extra paddles, rope, t-shirts and anything else to hand. Our masts looked more like scarecrows on the water but we hoped that if a crocodile did launch itself out of the water at us, a mouthful of scarecrow would be enough to

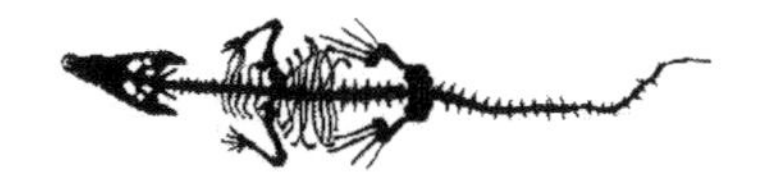

make him release and give us a get-away chance. Perhaps this was high hoping, but regardless, the scarecrows made us feel a good deal safer.

The days were long and the river banks dry and desperate for rain. There were towns and small river tributaries but the landscape was parched and brown. Here again we were attacked by a croc.

I was in a small section of braiding running parallel to where Ace and Rei were. I was not paddling and was sitting quietly in my kayak with my hat over my eyes having a quick mid-paddle nap, when I felt a rumble of scales beneath the kayak.

My breath caught. I knew exactly what that was. I opened my eyes to see a large crocodile sliding out from beneath me! Had he had been testing what I was, using the receptors in his back? Would he attack?

But sometimes you cannot control your reaction. As the end of his tail came out I grabbed my paddle and took two quick back strokes away from him.

The croc sensed the movement and turned back on himself, launching out of the water; he missed my scarecrow by a few inches. My frenzied paddling gave me a body length on him but he must not have really wanted me for lunch, for although he chased me for a good eighty metres or so he gave up the chase.

I did not stop my frantic paddling! With no less energy I paddled hard and caught up with Rei and Ace panting heavily.

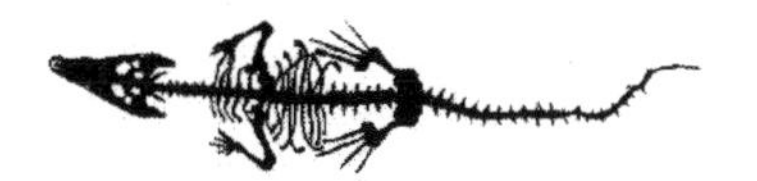

'We need to re-assess and re-vamp our scarecrows boys. I've just been scoped out!' I relayed the events and we pulled up to re-think out croc strategy. Higher masts, more flapping pieces of material .... Would it work?

All along the river here, the locals had built stockades with sharpened sticks in U shapes off the banks, so as to create safe places in which to collect water and to wash. Obviously crocs were bad news here.

The wildlife had thinned out now and even evidence of poaching had disappeared. There were still a lot of locals with dogs, probably hunting the last remains of wildlife amongst the maize, golden and dry now.

Carmine bee-eaters darted above us, open-billed storks collected in rabbles and one evening we caught a very small bull-shark. This was surprising for us; we did not know that bull sharks could tolerate fresh water or were we closer to the ocean than we thought? We could feel its pull and the shark only lured us on. Just one more river-bend, we thought, just one more.

Of course there was still a huge distance to go; over one hundred kilometres. A week or more of paddling.

The weather was more muggy now as there was more cloud cover and we wondered if we might get rain; we had had very little the entire trip. Here and there were small tourist camps - some even hunting camps. We pulled in at one camp and the manager showed us the collection of gin traps and

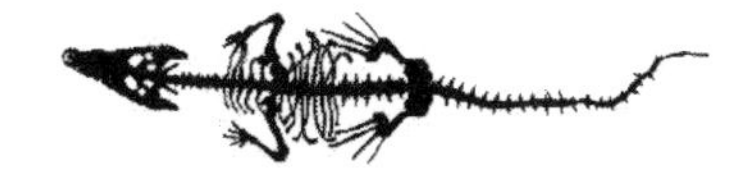

snares that he had collected in and around the area. It was sad to see how wildlife-poor this region was.

Local canoes acted as ferries across the river and were often loaded to danger point. But still the people smiled and waved. Mawire had given us each a Frelimo shirt and we wore them now, the bright red signalling us out on the water. The folks in the water ferries loved that we wore them and broke into whoops and Frelimo signals.

In most places the river was just over a kilometre wide, broken into braids, but now we stayed closer together in anticipation of more croc attacks. The closer we got to Caia, the more channels there seemed to be. Although we tried not to get distracted from the main channel we found them interesting. Some were river tributaries that ran parallel to the main flow for kilometres before they broke into the Zambezi again; others were possibly dominant flow channels from yesteryear. We had hit the estuary floodplains, even as far back as this. The dams above had full control over the fresh water flow down here and the rich sediment was less evident. Bank erosion meant that the water became a muddy brown, and at night the mosquitoes were worse than ever.

Sometimes we were curt with each other – just waiting for the end that was in sight. Sometimes we found every joke or action funny as we desperately attempted to make time speed up. We were ready for home.

But then the excitement and the interest would surge back.

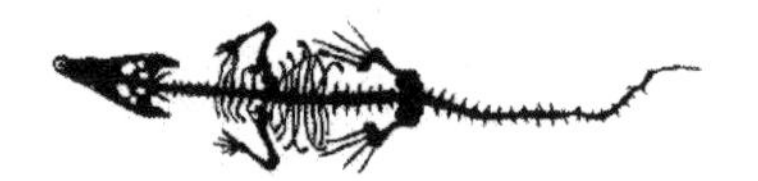

At Vila de Sena we came upon the impressive Dana Ana railway bridge. This bridge is over three and a half kilometres long and it's curved steel sides numbered over forty, and only in May, five months previously, had it been re-opened as a railway bridge at the cost of seven million US dollars.

Long sand islands lay like discarded lumps of play-dough in the river now, and we navigated around them, heading downstream to Caia. Here the vegetation was greener and more verdant. All the islands though had been decimated. Wildlife was nowhere to be seen.

At Caia we were met by a giant concrete structure with suspension wires. The bridge was almost new and was a feat of engineering at sixteen metres wide and two and a half kilometres long. The bridge cost around eighty million US dollars. The region of Caia has an AIDS epidemic that has attracted aid from all over the world. It also has an airport with nine hundred metres of paved runway.

But more than these Caia held no interest for us and so we paddled straight through.

Not far after the bridge at Caia as we paddled sedately in a formation that held Rei a good twenty metres in front of Ace and I, we saw Rei go into full panic mode. Neither Ace nor I could see any croc but we knew there must be one!

We quickly paddled closer and spotted a large crocodile, its sights set on Rei.

'Paddle!' I shouted quite unnecessarily - Rei was already paddling as hard as he could towards a

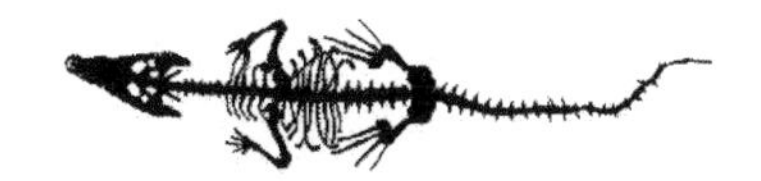

low mud bank on river right. In his panic he hit the bank full on and the current turned his kayak into the face of the oncoming croc.

We too paddled hard but Ace and I could do nothing - we were too far back!

Reilly though was quick, in the moment of contact he leap-frogged out of his kayak and onto the bank. He was safe!

Moments later Ace and I pulled up next to Rei and together we shouted and threw river debris at the croc until it had been scared away. Rei was badly shaken and together we built a fire and brewed some tea. What better way to recover from a croc attack!

Every day we would have tea at least three or four times. When tea called we simply pulled up to the closest shady tree, collected enough kindling to boil the kettle and filled up there and then. The closer we drew to the end of our adventure the more exhaustion seemed to creep along out veins. Mentally we flagged and physically we sagged. Our spirits were still strong though, and we pushed on, paddle stroke after paddle stroke, looking frequently at our maps and measuring the distance to the ocean.

'How far now?' we'd ask each morning as we awoke. 'How many paddle strokes left? We must have dipped our paddles at least a million times!'

'Not far,' was always the answer. We would take turns at guessing what the ocean would look like when we got there, if there would be anyone

there to meet us and how long we would stay. Spur winged Geese whistled beside us and on one sandbank stood six pink flamingos.

Here and there small churches and mosques clustered. Specifically we were looking out for a landmark that would allow us to find the small church in whose graveyard Mary Livingstone née Moffat lay. Julie Davidson (a travel writer for the Telegraph) says:

'*Mary Livingstone is a whisper in the thunderclap of her husband's reputation. Yet her own feats as a traveller in uncharted Africa are unique. For five years she toiled and taught at Kolobeng, the only white woman at the most remote mission station of the London Missionary Society. She was the first white woman to cross the Kalahari, which she did twice, at great risk to herself and her children; she was the first white woman to reach the Chobe river, part of the water system of the upper Zambezi valley.*'

At the river's edge there was a huge clump of bamboo, the only bamboo that we saw on our journey, and this, we had been told, was the landmark for us to climb up the river bank.

Up from the river lay the crumbling cemetery of a Catholic mission in Chupanga, which had been destroyed in the civil war.

'Here boys, found it' Ace called as we wandered in scruffy river clothes about the small and equally scruffy cemetery.

Mary's grave was lonely and neglected with

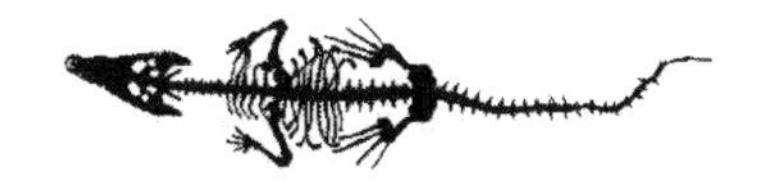

only a long-since desiccated bunch of flowers at its headstone:

'Here repose the mortal remains of Mary Moffat, the beloved wife of Doctor Livingstone, in humble hope of a joyful resurrection by our saviour Jesus Christ.' Mary had died on the 27 April 1862.

We paddled away from the graveyard in quiet respect and that night we mistakenly camped next to some vines called buffalo beans. Each plant has tiny little hairs that itch madly.

'Jamie why you scratching so hard?' Ace called across a smoky fire.

'Something itching' I replied.

Soon we were all frantically itching and cursing.

'Right I'm done with all this!' Rei finally spat. 'I'm going to bed.'

But even our sleeping bags itched. Even more.

We leapt up and scanned the water for crocs then dived in. *Scratch, scratch, scratch. What the hell was itching so much?*

Morning brought the answer.

'Look' Ace shook his head and we looked at his feet. 'Buffalo beans.'

Oh for the ocean! We wished hard on the stars that night. There were seventy kilometres left

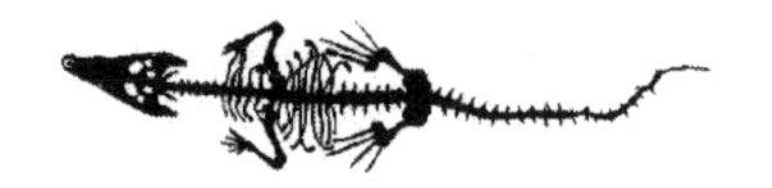

to go.

Soon the sugar plantations of the river mouth started. Huge, filthy barges moved the sugar up to the factories and the sugar blocks down to the ocean port at Beira where it was exported or shipped up the coastline. A place called Marromeu was the head office and here we wondered if a message might have been left for us.

We stopped to ask. In Marromeu, a local club had been built for the employees and we found cold beers on tap. There we no messages but there was a phone so we rang home and made a plan for John and Judy and Stu Reid to meet us here at the club after we had made the ocean. We would hit the Indian Ocean and catch a barge back up to Marromeu.

And so we left Marromeu with the end of our adventure in sight, and with the knowledge that the community had decided to slaughter us a pig in celebration of our finish.

Later the next day we saw our first coconut tree but the water was still fresh, it had not yet started getting brackish and the crocodiles were still very visible. We were very vigilant now that the end was in sight and avoided making any silly moves.

The last days fell into a pattern of their own. We left camp at first light and would rest in the heat of the day and then paddle till late in the afternoon. We stuck to the northern banks where the river channels were not *too* hard to follow and now the water began to taste salty.

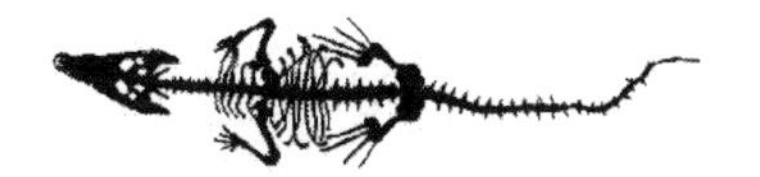

Belts of mangroves began and there was very little accessible dry land for us to camp on. The bird life increased and changed dramatically; there were sandpipers and plovers and large goliath herons intent on their fishing. Pigeon numbers seemed to grow exponentially as we neared the ocean and allowed us some sleep on comfortably full bellies.

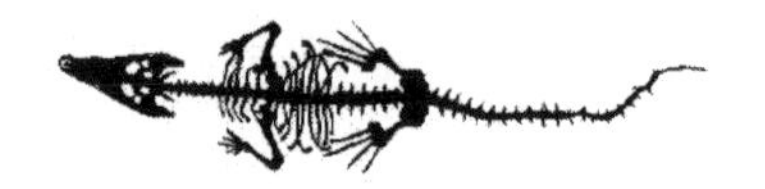

Skewered Rock Pigeon

A Jamie Oliver favourite!

Alternatively stuffed pigeon cooked slowly amongst hot coals is as delicious.

Pigeon is one of the most delicious meats there is. It can be cooked in numerous different ways, but out on the river the best is bbq'd and then wrapped in foil – foil is something every expedition should have plenty of.

Serves 3 hungry bird catching kayakers

3 x long green sticks
2 x spotted rock pigeons each
Onion (if possible)
Spices (garlic, tarragon, thyme, cloves, ginger, nutmeg)
Chilli (oil)
Wild figs
Sugar or honey

1. Pluck and gut the pigeons and wash them in the river.
2. If you have onions, peel and roast them ready to skewer.
3. Cover the pigeons with oil and crushed garlic and all the spices you have.
4. Skewer the pigeons on a green stick each.
5. Cook over hot coals until nicely browned.
6. Add figs and sugar or honey and wrap in foil and cover over with hot coals.
7. Leave to cook for ten minutes.
8. Add chilli and salt to taste.

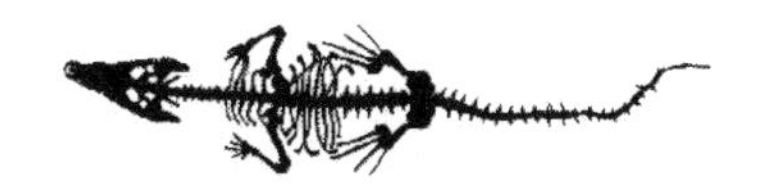

And then it finally happened .... One morning as we pored over the maps, we realised that *this* was the last day. Today we would see the ocean!

On the 25th October 2009 we hit the Indian Ocean. We had reached the end of our journey. My heart felt so full that it reached up and shook my head all on its own.

We had made it, we were here. This *was* the end!

As we drew to the close, we could see the old weather-beaten smokestacks of a sunken paddle boat that used to ply these waters. We could see the waves, small crashing white ponies. The long purple sea raced towards us and thundered against the shore. The palm trees seemed to rock-n-roll in the breeze and the spray.

'At the beach we celebrated. We ran up and down in the sand, we stripped off all our clothes (as celebrating boys do) and dived into the waves. We photographed each of our boats and us with them, in groups and on our own, and we poured our water that Judy had bottled for ea each at the source into the murky brown waters of the Indian Ocean.'

There was an immense sense of satisfaction for ourselves and for us as a team, both for what we

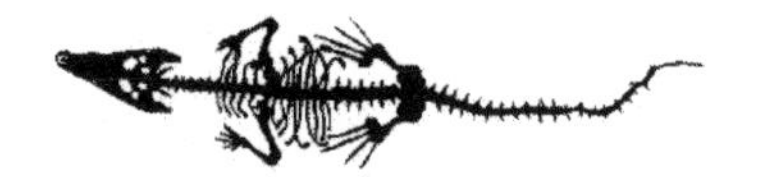

had achieved and for what lay ahead. It was exhilarating that we had accomplished what we had set out to do, and we felt lucky that we had made it against all odds; three non-kayakers against two thousand five hundred kilometres of river!

That night we camped on the beach with a big fire and lots of stories, only the whiskey was missing. The stars twinkled above us and we revelled in our success and rubbed our blistered and calloused hands together. As I bedded down in my filthy sleeping bag I could hear the roar of the ocean in my ears as if it were singing.

After a time I felt that I could never have been complete without all that had happened, and with that, sleep climbed in and settled comfortably across my features with a smile.

The very next morning we wandered up the beach to Chinde and celebrated with a few beers. Electioneering was in full flow and we wore our Frelimo shirts just in case. The town was in a sad state of disrepair and the once colourful, bold buildings were faded and tired, the roads potholed and littered. Many of the buildings were no longer used and lay empty – an ice cream shop, the old cinema, a garage - all were abandoned. There were avenues of trees against beach boulevards that had no traffic at all.

As we sat, cold beers in hand, at a small local eatery three police officers approached.

'Ummm.' We looked at each other slightly worried. *What now?*

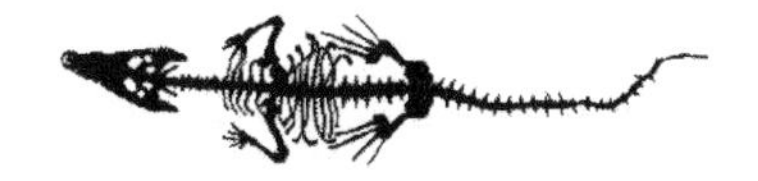

One uniform took out a small notebook and pen and cleared his throat threateningly. 'We must know if those plastic craft on the beach belong to you' he asked.

Ace informed him they did.

'Yes,' the uniform said 'us people do not move with such modes of transport.' He paused and cleared his throat again. 'These crafts are not licensed modes of transport in Mozambique and without a licence they are illegally with our country!'

We smiled together.

''No problem officer,' Reilly said 'I'm sure you are able to licence our kayaks for us.'

The officer and his uniformed back-ups were only too willing to sell us a licence and share a beer while on duty.

And so at the very end of our trip we licensed our kayaks and were legal!

Now we had to find a barge back up-river to Marromeu sugar estate. Who better to ask than the policemen? They were thrilled to help and used their uniforms to order us a place on a barge that was leaving late the next afternoon. It was a day behind our schedule but we had no choice.

As the tide scurried up the beach and into the Zambezi we packed our gear and loaded our kayaks into the barge. We strung up our hammocks, pulled out the *mascottis* and opened cold beers as we prepared for the eleven hour barge trip.

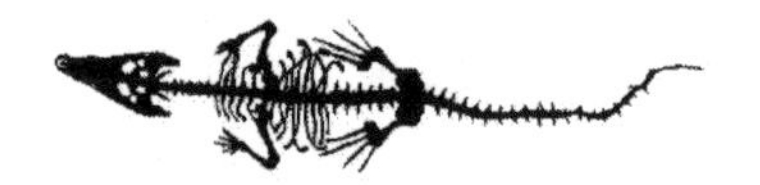

The people of Marromeu and our families met us in the full throws of a party with delight and beers, our promised hog roast had been eaten the day before, and there were no remains!

We had missed our own party! But there were hugs and kisses, exclamations and smiles all round before Candice asked Reilly for the promised Bible.

'Sure thing' Rei said. 'Let me get it now.' Reilly hunted through his dirty mess of equipment and pulled his Bible out. It was in a plastic bag and wrapped in a colourful sarong.

'Here you go,' he smiled triumphantly.

'Thanks Rei.' Candice removed it from its bag and unwound the sarong from it. She ran her fingers over its cheap hard cover, looked up at Rei and then opened it randomly. Of course it opened at one of the many missing sections.

Candice froze, 'Rei, why are there pages missing?'

Ace and I looked at each other – realising that we never did quite get around to telling Rei that we had been smoking his Bible.

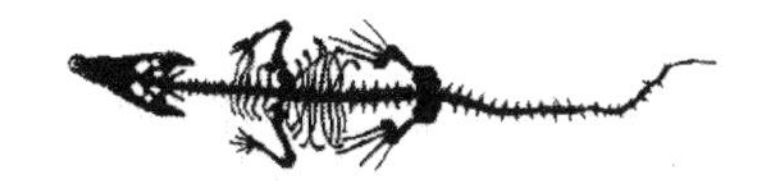

The End.

Don't forget to look at our blog where you will find a link to the online version of the

COFFEE TABLE PHOTO BOOK

https://blackmambaforbreakfast.blogspot.com

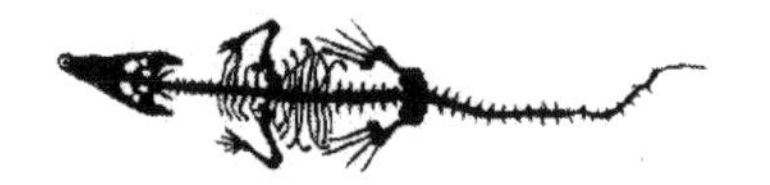

Take Away Tips For

Expeditioners

As you have seen our organisation of this expedition was exemplary and our advice really should be a pay service ... you don't agree? Well have fun reading it anyway!!!

Why should you Expedition? Expeditions are real-time experiences that cannot be paralleled. Maybe it's a ten day expedition because you are an American with a job that does not allow too much holiday, or you are a banker in Canary Wharf and want a challenge in an environment that you have spent hours researching on the Internet (only when you have no work to do of course). Maybe it's an expedition that takes in all pubs on the west coast of Hawaii (surely it's been done), or an expedition around Australia smoking *bhangi* with people from all over (that's definitely been done!), or perhaps you just need to expedition so as you can re-assess life ... do it! Expeditions are not only for the specially trained ... they are for anyone with tenacity and a love for life.

Deciding where to go! While making your decision to explore, weigh yourself up as a person: Are you the 'first decent' guy (or gal), the 'truly wild and scary' type or the 'I just want an adventure' type?

Choosing your expedition mates! Think funny – you need a joker on your expedition. Think planner – you need someone who will go the distance in

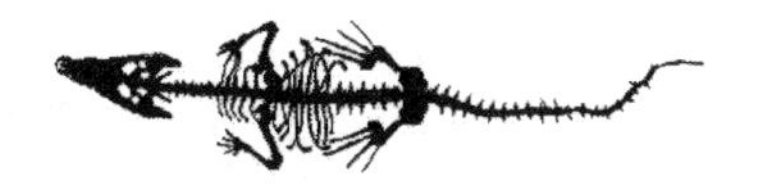

planning and research. Think realistic – is at least one of you medically able? Finally think friends – bonds are made and broken on expeditions, friendship is put to the ultimate test. (Oh, and if you are the philosophising sort who wants to learn more about your inner self ... choose wisely!)

To hire a guide or not to hire a guide? That is the question. If you need safety from marauding wildlife because you are ill prepared to deal with it – hire one! If you are up for a challenge and realise that you may not come home again – don't hire one!

Source-to-Sea! Hurray you've made a decision. Now is your team poetically inclined? Choose a beautiful meandering river in a first world country (the Mississippi). Is your team madly inclined? There is always the Congo or the Ruzizi River (remember Gustave?) Is your team considerately inclined? Choose a social river with good fishing (Russia in the summer). Is your team looking for an Ace Cracknell toughest expedition ever? Choose a river no-one has ever heard of in a country with an unstable government (Nile in South Sudan)... surely you have heard of the Nile?

Important Expedition Considerations

Beard Growing: (Ladies this should perhaps read 'shaving your legs!') *Hmmm* a tough one. If you are expeditioning in the tropics a beard keeps sunburn at bay but gets scratchy. It can however help in providing tasty morsels that escaped from the last meal (great attribute if hunting-gathering is not going too well).

Washing ... and smells: Let's just face it, if

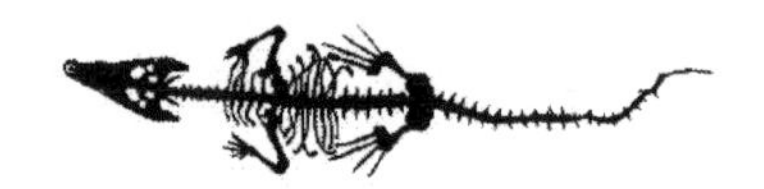

you are going to go on an expedition you are going to smell so you better just get used to it. If you are going to take soap along, take dettol soap – at least it keeps infection at bay!

Impressing people and collecting phone numbers: Note numbers 1 and 2 should *not* be adhered to if there is someone worth impressing (and believe you me, they appear from nowhere!) Be cleanly shaven (lads and ladies), wash well and have a few funny anecdotes at the ready – ones that make you impressive of course!

Creating an expeditionary blog: This is a rather new phenomenon but a wise move if you want people to know about your expedition; at least choose a catchy title (Surviving the Wanzilu ... with a warthog) and write engagingly.

Discuss writing a book ... then write one: I've read a lot of expedition books out there that are of such fantastic journeys but so hard to read (*oops* .. I hope you think you just read a good book!) If you can't write then hire someone who can. If your book is not good then no one will care about your expedition except your ever-forgiving family and friends. And if you can, hire a good illustrator (unless you are an incredible photographer – most people can 'armchair-expedition' with Google images quite happily!)

Happy Expeditioning

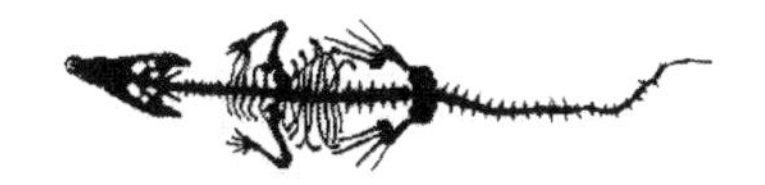

And if you are going on expedition soon then you might need to know

'How To'

How to ...find a honey supply

There are three ways you can do this:

The unconventional Way is to follow a lesser or greater honey guide. The honey guide will land nearby and will encourage you to follow it with a chattering and wing fluttering. It will fly a short distance and then return to see that you are indeed following it. It is important to ensure that a fresh comb is left out for the helpful honey guide.

The Easiest Way is to find a local who knows the area and who can show you where the honey is generally collected from. Then the honey must be shared between the kind local, the bees and yourself.

The Cheat's Way is to either ambush a local honey collector after you have stalked him to the honey source, or let him leave before you collect what remains in a hive that has already been raided. This is not a honey collection method to be proud of.

... deal with the bees

Africa is home to the killer bee and stings are both painful and can be dangerous. The hives are often found in tree hollows, up high in branches, in deep

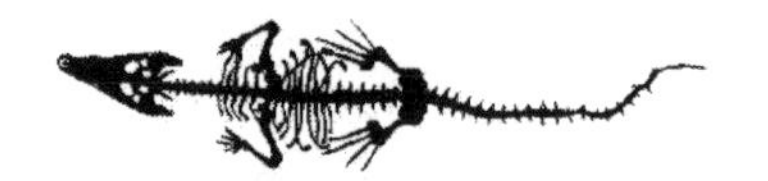

holes in stumps or in old termite mounds. Bees need to be dealt with calmly and efficiently with no noise. In fact they say bees respond to love, so emanate as much love as you can when you are in the vicinity of the hive and whatever you do, don't run!

...avoid being stung

If the honey is holed up, then smoke green grass under the hive. This smoke acts as a calmer and the bees become lethargic, allowing you to carefully reach in and take some combs of honey. It is preferable to collect honey at dawn or dusk or better still, in the darkness hours.

...use fresh honey

Fresh honey is extraordinarily valuable. It can be used to treat burns, can be used in cooking, in the place of sugar and can simply be eaten by the spoon.

...distinguish between honey types

Honey has a distinct taste that depends on which plants the bees are pollinating. *Acacia* honey is sweet and moorish, while bees that have fed on *Euphorbias* will make honey that is bitter and harsh. Before you collect honey, look around you and see which trees and shrubs are flowering. This will give you a clue as to the taste of the honey.

Honey is a poor man's gold so give it the full respect that it deserves, as well as its busy little makers. Without bees our world would be very different.

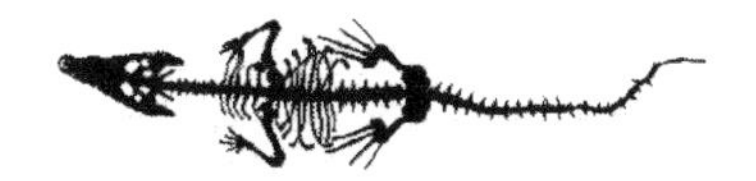

How to ... collect crocodile eggs

It's not as easy as finding a nest and raiding it. First look for the tell-tale signs of the crocodile on the sands along the river bank – drag marks from her tail with feet marks either side. Once you have found these you know that you are in the croc vicinity. Now look for regions in the sand that appear slightly damp, this will be where a clutch of eggs is buried.

Appoint someone on 'croc watch' while you dig down and carefully scoop out the sand to reach the eggs.

Never take all the eggs, leave at least two thirds of the clutch alone. Carefully cover up the eggs with the sand and pat it down – and get the hell outta there.

...light your campfire

There are three ways you can do this

The Conventional Way is to look around for a wide open space and to collect kindling, twigs and larger pieces of wood, in that order, to form a fire structure which you can stuff with dried leaves or grass. If there is good air circulation there will be good burning.

The Easiest Way is to use fire-starters. If you are the type of 'explorer' who has packed these then perhaps you should be warned not to get too far off the beaten track. That being said, if you have a leak-proof jar with some kerosene in it then you can add your dried out tea-bags and these are perfect fire-

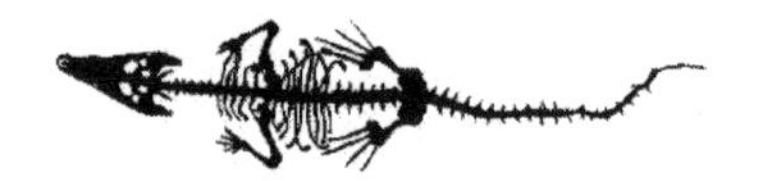

lighters in the rain.

The Cheats Way is to camp near a village and 'nab' some already glowing coals from a friendly local's fire. This method guarantees you'll have visitors that night and does mean that you have to keep to the lesser wilds.

...keep your fire under control

Don't let any sparks light dry grass around, and if your fire is big, make sure the flames won't lick any overhanging tree branches. Control the heat using different sized pieces of wood. When you have a nice pile of red coals, move these to the side to cook over and build up the fire again as light and warmth on the side.

...cook crocodile eggs nicely

Lay your frying pan on the hot coals and pour the whisked egg, herb and salt mixture into the pan. Let the underside of the omelette brown nicely and then flip the omelette carefully and do the same again to the other side.

Now and again shine your torch about the area in case there is a pair of glowing red crocodile eyes that are watching you prepare this lovely meal.

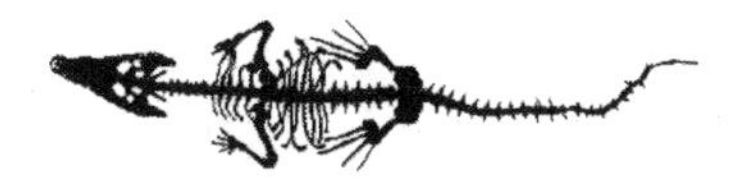

How to ... track a poacher

(Permission for use kindly given by the Bumi Hills Anti-Poaching Unit)

Tracking is indispensable to anti-poaching anywhere in the world. The better the tracker the easier it is to find and follow the poachers. The poachers know this of course and practice "anti-tracking" or "counter-tracking" measures to try and conceal their tracks or avoid leaving sign. Many poachers in Southern Africa were either guerrilla or counterinsurgency fighters or were taught by experts who fought in one or more of the many bush wars in Southern and Central Africa. They know how to stand behind a tree trunk to hide from aircraft, they know not to go near water points in the dry season, they know where to predict observation posts that have been set up to monitor them; poachers and thus their trackers must be patient, determined and skilled. Overcoming the poachers requires well developed tracking skills and a thorough understanding of counter tracking techniques. At Bumi Hills Anti-poaching Unit the trackers are trained in these advanced tactical tracking techniques.

The advantage is always with the poachers. If they know they are being tracked, they can easily lay an ambush on their own trail. Therefore tracking units try to follow without alerting the poachers.

... finding tracks

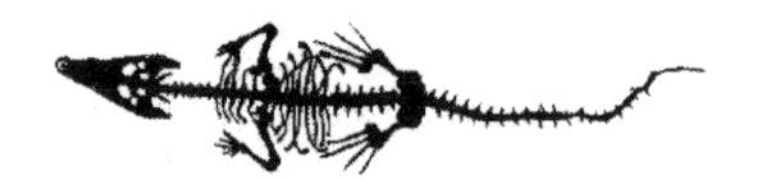

To pick up tracks the anti-poaching patrol units will 'cross-grain' areas where it is difficult to conceal tracks but necessary to cross, such as dry riverbeds, game trails, "capped" areas, watering holes and other sources of water. From here the freshness of tracks is determined and the decision to shadow or not made.

... tracking formation

The usual formation is a tracker with an armed scout on each of his flanks, who each move ahead of him. While the tracker focuses on following the tracks, the scouts focus on protecting against any threat from dangerous animals or ambush.

Gender: This is easy to determine when you know how. Women point their toes more inward and most important the straddle (the width between the line of tracks on the right and left feet) is much narrower than a man's. Simply, men walk with their feet further apart, whilst women walk with them closer together or even overlapping (picture a catwalk model walking down the ramp and a wrestler strutting in the ring).

Determining Stature: The height of a person is directly proportional to their foot length. Roughly 6.5 the length of a bare foot will give the height. This varies according to ethnicity and other factors.

Determining Weight: The width of the heel is greater proportionally to the length of the foot; the

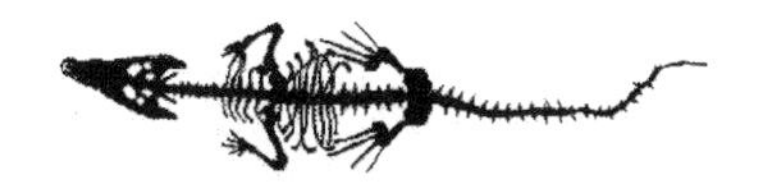

thinner the heel then the skinnier the owner and the thicker the heel the heavier the owner of the track.

Determining Whether Loads are Being Carried: When someone carries a heavy load they take shorter steps, they point their toes more outward and their straddle widens (they walk with their feet further apart). Furthermore, packs and other luggage will often be put down when resting and the signs left can tell us what exactly was being carried, i.e. box, water container, backpack, etc. Knowing how heavy a burden the poacher is carrying can tell how slow or fast they are able to travel or whether they will need to find water or not.

Ascertaining the weaponry being carried: Knowing what weapons and how many of them a group of poachers is carrying is crucial information. A couple of trackers simply cannot take on a large group armed with AK47s and RPG7s. As with other burdens, poachers will invariably rest the butts of their weapons on the ground when stopped. Every weapon is different and this ground imprint indicates the weapon type. A well organised and experienced group of professional poachers will often have one heavy calibre sporting rifle for shooting the elephants, and any number of assault rifles for use against wildlife protection personnel. Generally a heavy calibre .458 bolt-action rifle is designed to be used on big game such as elephant, while smaller calibre fully-automatic AK47's are designed for warfare.

Determining the Number of Poachers: This is

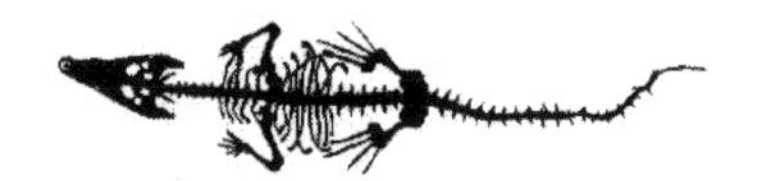

relatively simple. Once the direction of travel is determined two lines are drawn between the tracks furthest apart from each other. The number of people can easily be determined within the sectioned area.

Breaking Down the Group: Once the number of people is determined the trackers will assess the tracks of each individual thereby building up a picture of the make-up of the group and what equipment and supplies they have. For example, "serious" groups coming in from across the border in Zambia will travel in large, well-armed groups (they bring their own porters for the ivory), wear "takkies" (canvas plimsolls), carry all their water so that they do not have to go near the watering holes, and typically move faster. Local poachers on the other hand, travel in small groups because they can call on porters from local villages, they wear "manyatellas" (homemade shoes made from car tyres and tubes which leave very faint tracks) or go barefoot, they travel more slowly and carefully, counter-tracking to avoid detection. These groups often know where and when scouts will be, and therefore are less concerned about approaching water but will counter-track when doing so.

Counter-Tracking and Anti-Tracking: Experienced poaching groups use many methods to conceal their tracks or even to not leave any. Commonly this is done by not walking on ground that will leave tracks, such as stepping on stones, approaching roads, dry river beds and large game trails at a 45 degree angle and then leaving it at a

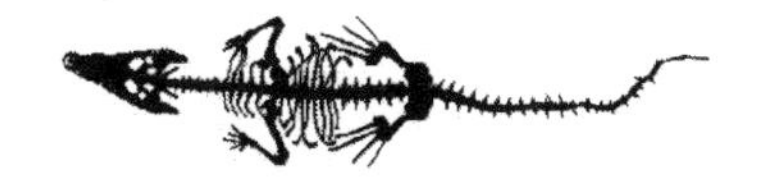

different angle after crossing, walking backwards across roads on one's toes, using the skin off the feet of baby elephants as well as many other tricks.

This is just a taste of what an anti-poaching tracker knows and does.

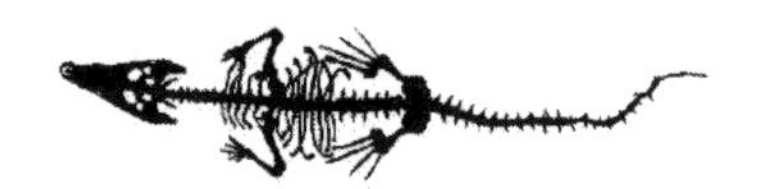

Epilogue

Africa has changed much in the last few years, even since 2009. There are plans for a new dam below Victoria Falls on the Zambezi. The Batoka Dam will do more to destroy tourism than anyone can imagine. The dam was first proposed in 1993 and will begin sixty-five kilometres down from Victoria Falls, at Moemba Falls. It will flood to within five kilometres of the base of the falls. This will effectively destroy the habitat of the rare Taita falcon that lives in the gorge, and will also affect the world class rafting and kayaking that this section of the Zambezi is famous for; one of the major draw-cards of the area. The proposed power station will supply up to 1,650MW of power, to be shared between Zambia and Zimbabwe. Electricity that is admittedly much needed to support growing populations.

Angola remains a country with many problems. Standards of living remain low, life expectancy and infant mortality rates remain very high and the majority of the nation's wealth remains concentrated within a very elite sector of the population. Angola remains the third largest producer of diamonds in Africa, having only explored up to 40% of its diamond rich territory. History has shown that countries in Africa with such potential natural wealth will be forced to continue the fight against corruption, human rights violations and smuggling.

Zambia is still a country that relies heavily on its copper industry despite the diverse base of rich resources. The government is now working to promote agriculture, tourism, gemstone mining and

hydro-power under which the Batoka scheme is listed. However still over 50% of Zambians live below the poverty line and the present rate of economic growth cannot support the rapid rate of population growth or the strain placed by HIV on the economy.

On August 4th 2013, Zimbabwe held its most recent election in and the tyrant Robert Mugabe won again. The country has hard years in front of it but will hopefully pull through with resilience. Mugabe has been in power now for thirty-three years and plans to serve another five year term. This will make him ninety four years old if he then steps down. The economical, social and political future is hazy.

Many believe that Mozambique continues to deal with high levels of corruption will do not help in a speedy recovery from its war torn days. Although the country is rich with extensive natural resources and has a growing economy that is based on food and beverages, chemical manufacturing, aluminium and petroleum production, it still suffers from inadequate infrastructure, commercial networks and investment. The next presidential election is scheduled for November 3013.

In all of these countries, Angola, Zambia, Zimbabwe and Mozambique, the Chinese government has won contracts to improve road and rail networks and is providing much needed investment. However, investment and infrastructure development come at a cost. Since the advent of the Chinese in Africa, poaching of elephant and rhino has increased drastically. There are no definite statistics that point towards a link but those on the

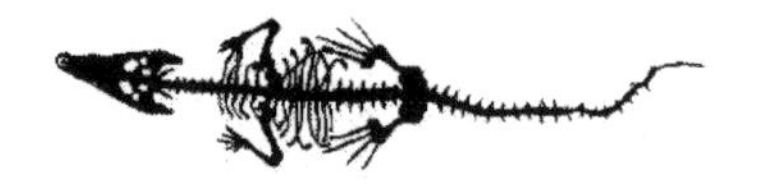

ground have evidence that much of the co-ordinated professional poaching is being funded by the Chinese.

Rei's grandfather, Norman Travers of Imire died in March of 2010. He was proud of the expedition that the boys had completed. Eddie Norris wrote later at the funeral, '*Only the elephants could have delivered such a moving tribute to Norman Travers. Shortly before he was buried last month on his farm, Imire, in eastern Zimbabwe, two 40-year-old bulls arrived unbidden, wandered through the crowd of 250 mourners, lumbered up to the coffin and sniffed it, long and intently.*'

Rei's grandmother and grandfather from his mum's side, Pat and Dave Hamilton, have also died and Reilly and his wife Candice have moved into their house (Welton House) on Imire and are expecting their first child in July 2017.

Sadly Mawire also left us of his own choice although his legacy still lives on. John Olivey left Imire and now lives in Harare running his own business. Justin Rogers who you met through his father Mr. Rogers who ran a crocodile farm and a concession near Cahora Bassa in Mozambique and lived under a tree, has finished school and is doing well. John and Judy are both still on Imire and three of their five children (Bruce, Kate and Reilly) live with them on the farm. Of the other two, Sam lives in the USA and Tara in Harare.

Tatenda died a sudden and unexpected death in August 2016 but he has sired a few young and the rhino population on Imire is growing. Kutanga, Pog and Tsotsi are also no longer with us.

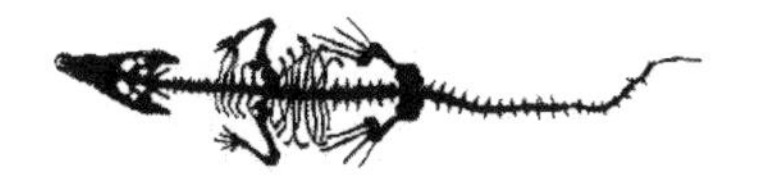

More and more work is being put into the protection of habitat so that the reintroduced rhino will be safe. Imire Safari Ranch continues to lead the conservation way in Zimbabwe, how long this will last depends on the road that Zimbabwean politics takes. But the Travers family will fight till the very end.

Poaching is decimating so many regions so that, perhaps by the time our children are born, there will be very few really wild places remaining. Poaching for elephant and rhino has soared this year (2017) and many parks services are allocating additional resources to fight the growing illegal trade of skins, bones and bush meat. The fight for wildlife preservation will continue for as long as there is a demand for products like horn and ivory – and while competition for land continues between man and wildlife. It remains to be seen if the rhino will ever flourish again.

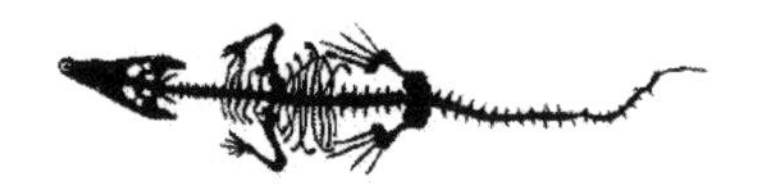

Appendices

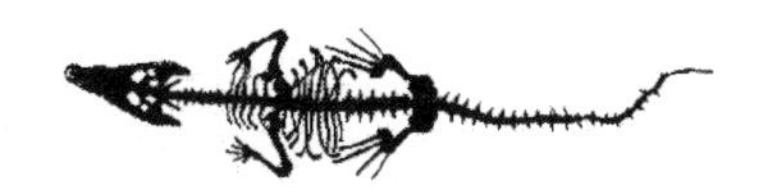

Row Rhino Row – Expedition Kit List

Foodstuffs

Posho (Maize meal)

Pasta/spaghetti

Packet soups

Nuts – all types

Tinned food

Bully beef

Milk power

Tea/coffee

Cigarettes

Muesli

Whiskey

Lemons/chillies

Spices

Sugar

Flour

Beer

Salt

Rice

Cooking stuffs

Gas cooker 4 cylinders

Cutlery/crockery

Matches/Lighter

Newspaper

Frying pan

Sufuria x 2

Teapot

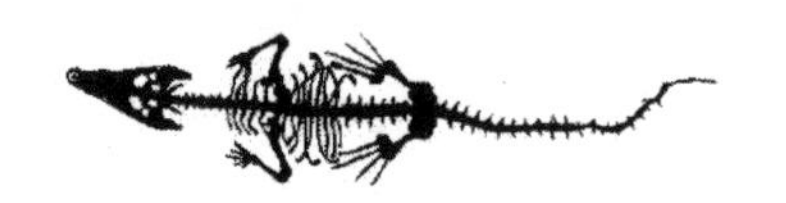

Sleeping bags

Mozzie nets

Hammock

Shit paper

Bhangi

Soap

First Aid

First aid bag must be waterproof

Water purification tabs

Vitamins

Scissors

Nutriaid for aids victims

Other

Gautex sleeping bag cover

Pencil sharpener

Bird/fish books

Passport holder

Essentials

GPS maps and compass

Binoculars/telesc ope

Lots of fishing kit

Leatherman

Fishing rods

Spear guns

Catapult

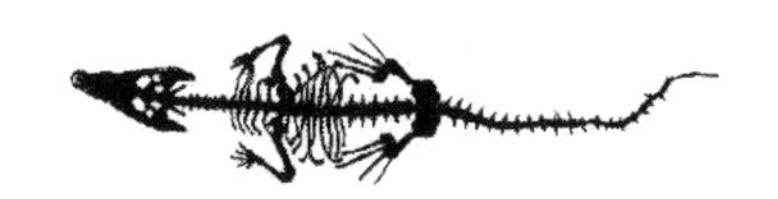

Panga/axe

Diary

Clothes

Sunglasses

Hats

Barber x 1

Kikoi x 2

Shirt x 3

Jumper x 1

Shorts x 3

Toolkit

Silicon tubes/wire/glue/marine weld/putty/needle and thread/strong line/duct tape/pencil flare/zip ties/tube 750 16s/aeroplane tube/JB weld/fibreglass/pv4 glue/plastic gunias

Torches with 2 x LED bulbs plus three sets of batteries

Waterproof cameras with spare SD cards

Solar panel with inverter and 12 volt battery

1 X 22mm rifle and 50-200 rounds

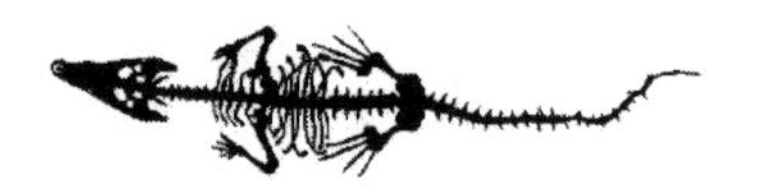

Dynamite for crocs

Knife sharpener

Sharp knives

70m rope

Spotlight

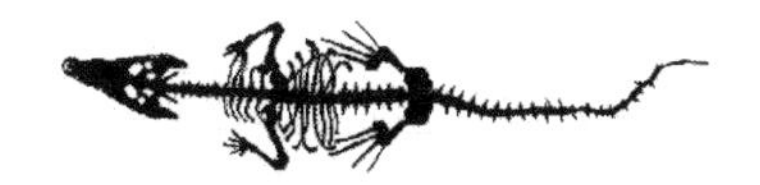

Further Information

For those who found the names and places of the people we visited interesting, here is some further information.

The Tashinga Initiative Foundation
Lynne Taylor
M +263 912 348 671
E lynne@thetashingainitiative.org
S ldtaylor1
www.thetashingainitiative.org

Imire: Rhino & Wildlife Conservation
John and Judy Travers
M +263 (0) 222 2354
E volunteering@imire.org
S imiresafariranch
www.imiresafariranch.co.zw

Bumi Hills Anti-poaching Unit
M (+263)772135665
E conservation@bumihills.com
S nicholas-milne or rodgers-matimbidzire
http://bumihillsapu.blogspot.com

International Anti-poaching Foundation
M +263 (0) 774 659 474
E Ace.slade@iapf.org or steve@iapf.org
Online www.iapf.org

This organisation fights poaching in volatile regions

George Adamson Wildlife Preservation Trust
M +44 (0)20 8343 4246
E info@mkomazi.de
S kora National Park

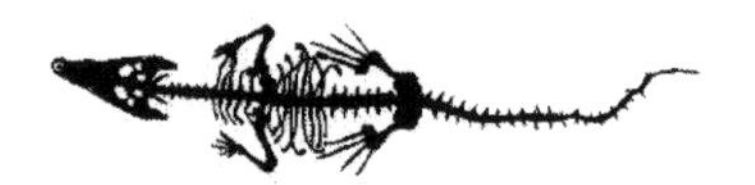

www.georgeadamson.org

Imire Volunteer Programme
Email: volunteering@imire.co.zw
Tel: +263 774 433 063

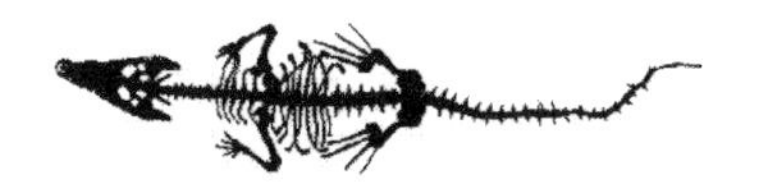

Sponsors

Thank you to all who helped us both before the expedition started and along the way. If we have not mentioned you, it is not because we have forgotten; it is just that there are too many people to name individually. Each of you knows who you are and our heartfelt thanks go out to you all, river-folk, bush-folk and city-folk. Below I have listed our main sponsors.

Northern/Bindura Haulage Ltd, thanks for the kayaks, the paddles, the PFD's, the shirts and the fuel that got us to the source and home again.

Olympus, thank you for the two tough waterproof cameras

Imire Safari Ranch, thanks for the biltong that you supplied throughout the expedition. And Judy and Peeps, thanks for the muesli. Thanks also for the fuel that got us to the source and home again.

Imire Volunteer Program, thanks for the one kayak that you sponsored and that I hope will give volunteers a good time on the dam for years to come

Mike Daines Design/Signs of the Times, thanks for our Row Rhino Row logo and for the printing of the stickers to go on our kayaks

Caterpillar, thanks for the hats and jackets

National Foods, thanks for the river food

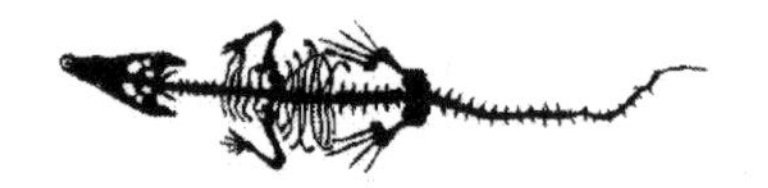

Glossary

Baas – Polite term for someone, originally meaning Boss

Bahati – Means 'lucky' in Kiswahili

Bhangi – Tobacco + +

Bitings – Our term for 'nibbles' or finger food

Boerewors – A type of South African sausage
Bokkie – An older man

Bollocking – Telling off

Bomas – These are stockades designed for animals

Bru - Brother

Catapult – A slingshot

Hundreds – Short for a hundred percent

Hodi – Kiswahili for 'hello, anyone home?'

Hotelli – Kiswahili name for a hotel

Karibu – Kiswahili for 'you are welcome'

Kopje – The Southern African term for a pile of balancing rocks. These can often be the size of small houses.

Kwacha – The Zambian currency

Leatherman – The brand name of a very popular pocket knife/pliers set

Lekker – Southern African term meaning 'good' or

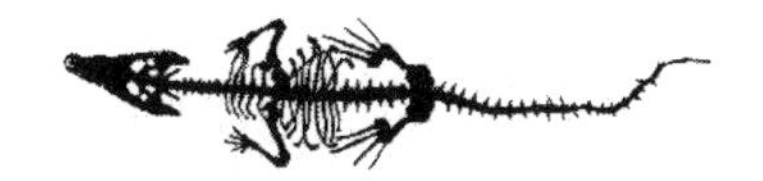

'cool.' Similar to awesome.

Lightie – Small child

Makuti – A thatch style of roof

Mascotti – Cigarette roll-ups (with a kick!)

M'ganga - Witchdoctor

Mitungi – Kiswahili for jerry can

Muti - Medicine

Mzee – Term of respect to an older man

Oakes – Southern African slang for 'a person'

Panga - Machete

Penga Penga – Slightly crazy in the mind

Plonker – Term from 'Only Fools and Horses' that is popular, meaning 'you idiot' in a friendly manner

Punda wewe – Kiswahili for 'you donkey!' A term used by myself often, like 'plonker.'

Rafiki – Means 'friend' in Kiswahili

Rhodesian Ridgebacks – A self coined term referring to people who are old enough to have lived in Rhodesia.

Shuppering – (Pronounced *shoo-pering)* Herding i.e. I *shuppered* Rei along as we were out of time

Sufuria – Kiswahili for saucepan

Tembo – Kiswahili for elephant

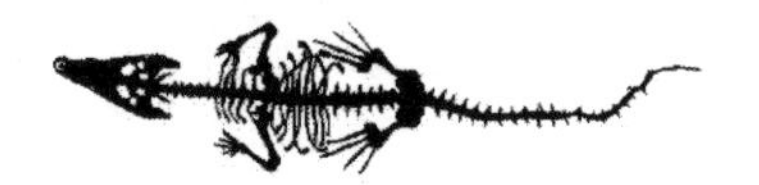

Vundu - Catfish

Yakanaka – 'Yeee-ha'

Wasi-wasi – Kiswahili term for slightly crazy

Zimbo – Slang for a Zimbabwean

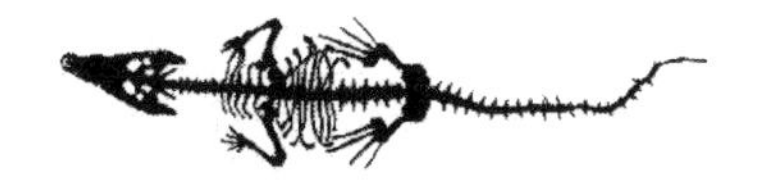

Co-Author's Note

'Jamie you should write a book about your Zambezi trip' I said.

'Holy smokes, you must be joking! Me write a book? No way,' he said. 'You do it.'

And so in the two months that I spent with him up at Kora National Park in Northern Kenya, where he was an Honorary Park Warden working with rangers, we compiled and recorded the main events of the expedition. Later he journeyed to Uganda and together we sat on an island in the White Nile and revised and edited until we were happy.

It was not possible to kayak the entire Zambezi in 2009 when he tried, but *Black Mamba for Breakfast* is the true story of the parts of the river that he did manage to see, from the source to the sea, as told from the perspective of Jamie Manuel. This story is drawn from the memories of Reilly, Ace and Jamie; from their diaries, fire-side stories, as well as from dozens of hours of interviews.

To weave the stories from the three boys into one cohesive narrative I have been forced to omit some anecdotes, and to move others to dates or locations that vary from the times or places that they actually occurred. In real life, events happen that relate to each other in ways that cannot always be described easily. In writing; one has to arrange these connections in a more easily readable format. However, no incidents have been made up and I have told the major events of the story as accurately as I can.

It has been necessary to reconstruct dialogue: something that is ever fraught with

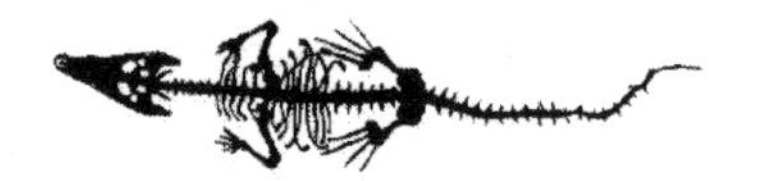

difficulty, but as I know my brother and Reilly well, I feel that I have been able to hear them throughout the writing of this book. I have not had the chance to get to know Ace so well and so have had plenty of help from all the boys in remembering conversations.

Memory is inherently subjective, and some of what Jamie remembers, the others contest. In these cases we worked to achieve a compromise. However, as all the boys kept diaries throughout, in most cases the disagreements were only slight.

I have used some non-English words throughout the text and have included a glossary at the back to help with this. It is common among people who have grown up speaking English in an African country, to pepper their speech with words from one or more of the many tribal languages within their country. In Zimbabwe these are KiShona and IsiNdebele and in Kenya, Swahili; the boys used such phrases often, even twisting the words into phrases that became uniquely theirs. Nowhere in this book have I added fictional events or characters.

Everyone whose name appears in this book is a real person.

The Zambezi expedition forms only the main thread of the book. I have used it as a link for all of the other stories that have captivated not only me over the years, but all others who have heard their telling too. Again, I have met and know all of the people who populate these 'bush tales' and believe that I have managed to capture their essence in dialogue and in description.

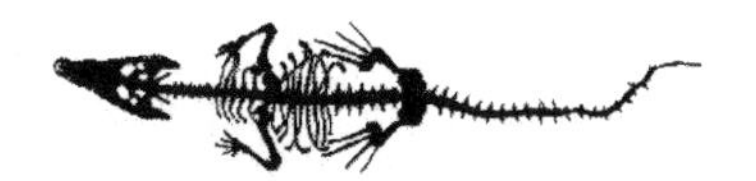

The people whose stories fill this book are an amazing bunch. They may be slightly 'off their rockers' or, 'in another world' but they are real and many of them live and work in extremely tough conditions and deserve a huge amount of respect.

Hollie M'gog, December 2013

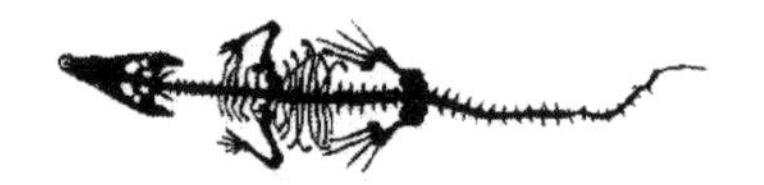

Acknowledgements

This memoir stroke adventure-travelogue is ghost written by my sister Hollie, and to her I am grateful for the long hours she listened to and recorded my stories, so that she could re-listen tirelessly, again and again. I am grateful to her for the hours she put into reading my handwriting in the cheap cardboard diaries that we all kept, while Reilly and Ace allowed her to interview them extensively and to extort from them all the less legal memories of the trip - (Ace you weren't so easy to get hold of)!

Also deeply appreciated are the sponsors that put money forward and allowed our dream to become a reality, even if we were ill-prepared and not fully aware of what we were setting ourselves up for; their names are listed in the appendices.

Enormous thanks must go to Reilly and Ace. Together we shared one of the most rewarding and difficult experiences in my life thus far. I could not think of two better people to have shared such an adventure with.

The biggest thanks, though, must go to Africa. For it was she who provided the most fantastic of all landscapes for us to explore, to hunt in, live in and dream in. It was she who spawned the mighty Zambezi and gave us incentive to undertake such an expedition.

We thank all those who generously shared their stories, experience and skills with us, and without whom the expedition may not have been

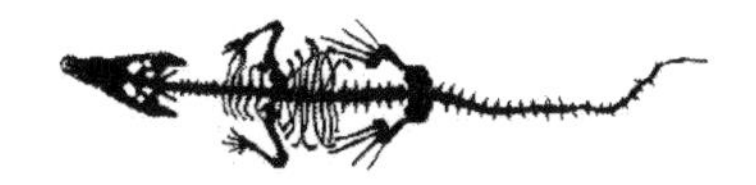

possible; kayakers in Vic Falls, old Africa hands, hunters, farmers and wardens. Keep up your good work and remember that without you all, Africa would not be as spicy as it is.

Thank you everyone for your belief and courage in us; John and Judy, Stu and Peeps, Mum and Dad – you all believed, and for that we will be ever grateful.

Jamie Manuel
Kampi Ya Simba
July 2013

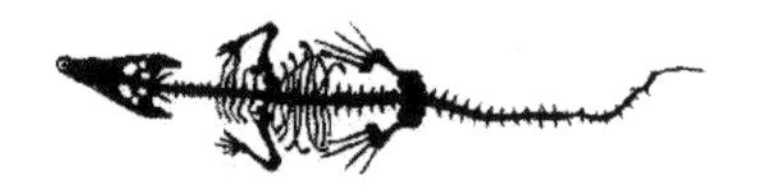

Don't forget to look at our blog where you will find a link to the online version of the

COFFEE TABLE PHOTO BOOK

https://blackmambaforbreakfast.blogspot.com

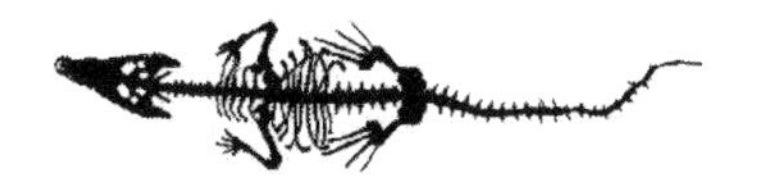

Author Biography

Hollie M'gog (Manuel), was born in Nakuru, Kenya in 1984 and although Kenyan, has lost her heart to Uganda.

Lucky enough to have been schooled on three continents (Africa, Australia and Europe) she has led a wanderlust life and has had many an exciting adventure herself, including being pushed sideways by a mountain gorilla in Uganda, having her hat lifted off her head by a charging wild elephant in Botswana, bitten and scarred by a wild orang-utan in the forests in Sumatra and slipping into a crocodile feeding pond in Australia when the embankment collapsed!

Africa runs in her blood stream and although she enjoys to travel, by way of kayaking, hiking and hitch-hiking, through Central and South East Asia, Australia, the West Indies and Canada, she will never leave Africa, it's colourful people, it's aggravating bureaucracy, it's intense wilderness and it's freedom from the 'general rules of society'.

'Black Mamba for Breakfast' is her first non-fiction book but she hopes there will be many more as she lends herself out as a ghost writer in order to vicariously live the wild adventures of others. She is currently working on a fiction Trilogy (Boda Boda Tales from Uganda: 'Long Rains', 'Short Rains' and 'A Long Dry Season.'

mgogwriting.blogspot.com

Made in the USA
Lexington, KY
09 August 2017